NETWORKED CONTENT ANALYSIS

THE CASE OF CLIMATE CHANGE

SABINE NIEDERER

Theory on Demand #31
Networked Content Analysis: The Case of Climate Change

Author: Sabine Niederer

Foreword: Klaus Krippendorff

Editing: Rachel O'Reilly
Visualizations: Carlo de Gaetano
Production: Sepp Eckenhaussen
Cover design: Katja van Stiphout

Supported by the Amsterdam University of Applied Sciences, Faculty of Digital Media and Creative Industries.

Published by the Institute of Network Cultures, Amsterdam, 2019
ISBN: 978-94-92302-42-7

Contact
Institute of Network Cultures
Phone: +3120 5951865
Email: info@networkcultures.org
Web: http://www.networkcultures.org

institute of
network cultures

CONTENTS

FOREWORD

KLAUS KRIPPENDORFF

Communication scholarship was born at a time radio and television became a challenge to professional newspaper journalists and the emergence of novel theories of communication. While it borrowed investigative methods from existing disciplines – experiments from psychology, surveys from sociology, ethnography from anthropology, and last if not least took advantage of what the new communication technologies had to offer – it contributed three major methods. One was digitalization, which grew out of information theory and ushered in our developing communication infrastructure. The second was content analysis, the systematic study of what was communicated, largely by the media to the public. And the third was the idea of networks, who talks to whom, and what are the social and individual consequences of complex connections. Sabine Niederer's Networked Content Analysis draws on all three indigenous contributions of communication scholarship.

From its beginning, content analysis aimed at making unobtrusive inferences from texts to their context of use. Its ability to analyze bodies of texts larger than what any one analyst could read and interpret called for methodological precautions not typical in literary scholarship. In return, it revealed novel insights not available with smaller data: historical trends, comparisons across different sources, and support for theories not recognizable by unaided scholars, as Niederer shows. It was also adopted by numerous other disciplines concerned with phenomena that are constituted in linguistic communication. While the 'content' that content analysis claimed to study remained metaphorical, often presented in terms of frequencies, this ambiguity invited statistical accounts not ordinarily encountered by informed readings of texts.

For example, contingency analysis charts the proximities of selected concepts in various communications. Co-occurrences in texts were shown to correlate with authors and readers' associations, manifest in their ability to recall them easily. Finding patterns of above and below chance contingencies provided a basis for inferences about the conceptual structures of individual authors as well as widely shared political, social, and cultural beliefs. These inferences were basically of a cognitive nature. Search engines vastly expanded the ability to discover co-occurrences in documents with three caveats: Search engines find strings of characters, words or phrases, not logically connected concepts. They are often insensitive of unequal proximities in documents, and when searching larger databases, leave somewhat uncertain what accounts for evident cooccurrences.

Tracing one authors' references to other works and theirs to still other works is another example of content analyses pursuing connections, across documents not within them. There are of course numerous reasons for citing other publications, but familiarity with their authors or their ideas underlies all of them. Citation analysis revealed not only a single authors' literary resources, but following the references of references could reveal how members of a discourse community hang together, the centrality of their individual contributions, where conceptions originate, and how diverse discourses influence each other. Citations are social acts and content analyses of citations offer considerable insights into how largely academic communities are organized and construct their objects.

Although the origin of the idea of hypertext has been traced to Ludwig Wittgenstein's hierarchical system of numbering of comments on propositions and comments on comments in his Tractatus (1922), it was not until the 1980s that digital texts enabled readers to click on links within a text to explore related matter, effectively enabling them to browse within a predefined textual universe. Hypertexts overcame the constraint of having to read text in the order it was written. It enabled readers to navigate their own paths through textual, visual, even auditory matter, towards their own intellectual goals. Content analyses of hypertexts had to chart the network of connections between the contexts of each link, a task that became quickly unmanageable without computational aids. Inferences from such networks are neither psychological nor social but have to do with the possible narratives one could extract from hypertexts.

Evidently, the recognition of networks of connections constructed in processes of analyzing the content of bodies of text has a long history. However, Niederer's Networked Content Analysis offers a quantum leap into the digital age.

Her work accepts the methodological premises of content analysis, appreciates its unobtrusive way of creating data, but adds tools and concepts to tackle the complexities of digitally available texts from Facebook, Twitter, blogs, websites, email, and electronic databases. While acknowledging the large volumes of online texts, to her credit, she is not letting herself be distracted by celebrating such volumes as some enthusiasts of quantification do, unrealistically believing that so-called 'big data' could identify large social problems with easily obtained statistically significant findings. Instead she explores such texts as the networked products of the socio-technological nature of diverse online platforms.

Niederer convincingly argues against treating social media as a mere alternative to one-way mass communications. For once, online texts rarely are single-authored and individually responded to. Their contents are the product of interactive collaborations, not only among individual contributors but also with diverse platforms that connect them. She argues that online content is platform-specific and accounts for their characteristics in terms of what she calls their 'technicities'. To make sense of online communications, she insists, users and content analysts must come to terms with these technicities albeit in very different ways. She adopts two guiding observations that networked content analysis has to acknowledge. The first is that web content is increasingly accessed and organized through the use of different search engines and platforms. The second is that the technicity of communication can no longer be separated from the analysis of networked content. While users of digital media develop and employ their own competencies, she argues that content analysts of digital texts need to acquire platform specific tools and literacies to recognize the dimensionalities, processes, and networks that different platforms facilitate. For instance, search engines provide search results in ranked lists, Wikipedia cleans and organizes multi-authored texts with robots, and Twitter links texts through hashtags of up to 140 characters in length.

A starter of networked content analysis is the use of computational methods to identify connections in large bodies of texts and generate visualizations of their multitudes, be they responses to tweets, references across documents, coinciding character strings, or links between websites. Such networks can be very complex and rarely lend themselves to simple narratives. For Niederer, visualizations of such networks can serve as navigational tools in conversations on how to proceed. To serve as such, analysts have to realize that such visualizations are the artifacts of the mapping algorithms that created them. To guide content analysts to answers to their research questions, these algorithms have to be compatible with the technicities that provided the analyzed texts. However, compatibility is not always demonstrable. For example, when the makeup of a platform is proprietary, as for Google's search engines, the content analyst is limited to describe their technicities in terms of their results, by what they do.

Niederer devotes one fascinating chapter on the technicity of Wikipedia, the collective, sometimes competitive editing of its entries, including by roaming editing robots that check on the grammar of its entries, eliminate inconsistencies, and most importantly, create hyperlinks among its entries as well as to outside literature. Users of the Wikipedia do not know who wrote an entry. The anonymity of authorship is part of Wikipedia's philosophy of remaining open to changes which is also held against its use as a quotable authority. Yet, the editing history of each Wikipedia entry can be examined by any user. It provides a sense of how an entry developed and how conceptual controversies are played out. Its entries evidently are organized by Wikipedia's technicity and their contents cannot be separated from its distinct operational features.

Aware of diverse technicities leads Niederer to qualify the effects of new media content. For example, while political scientists have described Twitter as a channel for mobilizing political actions, for example, leading to the 2010s Arab Spring, a pro-democratic revolutionary wave of demonstrations and protests in the Arab world, Niederer is more careful in describing Twitter as an awareness system that offers its contributors a sense of where they are within a particular technicity. When a tweet goes viral, she suggests, its popularity may not be the only explanation. Equally and perhaps even more important is its fitting the technicity of the platform that networks it. This interpretation is justified when examining the ultimate consequences of the Arab spring. To make a political difference requires other forms of organization not cast in 140 characters.

Niederer exemplifies networked content analysis by various applications to the debate of global climate change. Unlike traditional content analyses which tend to focus on biases in the form of unequal frequency distributions, explainable in psychological or sociological terms, the choice of a public controversy is well-suited to demonstrate its capabilities as online texts on a common themes include unlike actors advancing opposing arguments. To contextualize her exemplifications, Niederer situates the history and stakeholders in the climate change debate in the context of what public controversies consist of.

In the network extracted from Wikipedia entries, the choice of 'Global Warming' occupies a central node that is linked to numerous related issues, countries, economic issues, diseases, energy policies and scientific findings. Such a network can be looked at from numerous perspectives and be variously decomposed.

Realizing that the Wikipedia is naturally biased towards consistency, controversies can become manifest in the editorial changes of disputed features. So, one way of charting the heat of a controversy is by measuring the frequency of editing changes in climate-related entries. While individual editors are known only by their code names, Niederer describes algorithms to depict how many contribute to which entries and at which time, giving a sense of where the controversy takes place and the speed in which it moves. While it is not difficult to separate individual editors from robots, evidently the content of interest is inextricably tied to Wikipedia's technicity.

Attending to a different technicity, Niederer also explores algorithms to identify networks within large bodies of tweets. Co-hashtag analysis resembles contingency analysis but reveals connections among clusters of similar tweets. The inferences she draws from these networks deal with issues of the vulnerability of different themes (areas in the world, phenomena, and actors) in the presence of threats (food, floods, diseases, and weather), issues of how skepticisms and conflicts migrate from one cluster to another, how clusters adapt over time. Evidently these conceptions are based on the texts used in tweets but inextricably linked to the nature of the Twitter platform that individuals learn to navigate and content analysts need to acknowledge to make sense of these data.

Sabine Niederer's work responds to the changes that digital technology generally and diverse platforms for communicating among people in particular have introduced into our social world. Mass communication was a simple technicity. Contemporary communication is essentially networked. We create texts not just for particular addressees, but selectively rehash, redistribute, copy, and modify texts without being always cognizant what they do. The essence of online communication is no longer what is said but the networks we implicitly create, sustain, reawaken, or let go of. Networked content analysis begins to recognize the socio-technological infrastructure of our contemporary existence. It is a fascinating step into the future and well worth taking seriously and develop.

References

D. Stern. 'The University of Iowa Tractatus Map'. Iowa UP, 1996, http://tractatus.lib.uiowa.edu/.

L. Wittgenstein. *Tractatus Logico-Philosophicus* (New York: Harcourt, Brace & Co, 1922).

1. INTRODUCTION

This book has its origins in a project developed during the Digital Methods Summer School of 2007, the first annual summer program on methods and tools for social research with the web at the University of Amsterdam, titled 'New Objects of Study'. One week of the summer school was dedicated to 'Controversy Mapping - Citizen Equipment for Second-degree Objectivity' and the keynote speaker was the famous sociologist and philosopher Bruno Latour.[1] Via Skype, Latour provided an introduction to the mapping of controversies, based on the educational program he had developed at Sciences Po in Paris.[2] He started by outlining how to define and detect a 'good' controversy. A controversy is a 'shared uncertainty about facts', that manifests publicly through a range of attitudes. Latour includes consensus and agreement among the attitudes surrounding a controversy, and considers consensus an extreme moment in a controversy when actors abandon the controversy or agree.

Controversies can form and develop through hot arguments or cool disputes, depending on their intensity and the relative numbers of positions in disagreement over certain time periods. There is no such thing as a solid or fixed state of any controversy, or, for that matter, of consensus. Consequential to this temporal definition and its appropriateness to scholars' ongoing relation to controversy as a *research object,* and as a specific kind of *research practice,* Latour suggested that researchers should best be prepared to jump right into the middle of a controversy and describe what they encounter there. A 'good' controversy (i.e., a controversy most suitable for analysis) takes place across heterogeneous sources (e.g., academic journals, newspapers), and includes people from different disciplines. This range of actors can be studied through their specific vocabulary (the so-called *actor language*). It matters significantly in approaching the research of a controversy as to whether it is 'live,' past or present, and how many people are involved (and how many of them are scientists). One should beware that some controversies may be too big to research, involve too many actors, or too many points of contestation (the example Latour gave here was that of genetic manip-ulation). In such cases, it is best to choose a sub-controversy from a larger one. Furthermore, Latour stressed that researchers should describe all these dynamics of a controversy *without translating* what they observe into a more common or analytically familiar language. Steering clear of predefined keywords and categories enables researchers to better 'follow the actors' and log actors' language, connections, and formats.[3]

In Latour's approach, the actors of a controversy may be found at a specific event or gathering, in a collection of writings, an e-mail exchange, and so on. For my first experiment with a con-

---

1   See also the summer school's wiki page: https://wiki.digitalmethods.net/Dmi/MappingControversies.
2   B. Latour, 'Mapping Controversies', presented at the Digital Methods Summer School, University of
      Amsterdam, Amsterdam, 2007.
3   Tommaso Venturini, working with Bruno Latour in the Controversy Mapping educational and research
      program of Sciences Po describes 'three commandments of observation': '1. You shall not restrain
      your observation to any single theory or methodology; 2. You shall observe from as many viewpoints as
      possible; 3. You shall listen to actors' voices more than to your own presumptions.' T. Venturini, 'Diving
      in Magma: How to Explore Controversies with Actor-network Theory', *Public Understanding of Science*
      19.3 (2009): 260.

troversy mapping research practice, which I conducted with Esther Weltevrede, we looked at animals most frequently depicted and mentioned in the climate change debate on the web.[4] Looking at three different online spaces: the news (accessed through Google News), the web (accessed through Google Web Search) and the blogosphere (accessed through Technorati, the dominant blogosphere search engine at the time), we created word and image clouds of those animals resonating most in the climate change debate. These 'issue animal' hierarchies proved distinct per space, and this was the case in the textual as well as in the image analysis. The web gave attention to a wide variety of endangered species, giving way to those affected by global warming as well as global cooling. The News favored the polar bear, and also presented a new animal: the cow, which is not so much affected by global warming but one of the causes, as cows emit methane. The blogosphere showed a strong preference for the polar bear too. But a closer look at the actual imagery revealed that many polar bear images were of people dressed up as polar animals during activist protests. This also explained the appearance of the dogs in the data set: the activists' pets taken along to protests. The study pointed out that each online content space had its own hierarchies and needed research approaches adapted to its specificities, a finding that was worth exploring further.

Climate Change as a Globally Encountered Controversy

During the summer school of 2008, I chose to pursue the study of the climate change controversy further . In March of the same year, the Heartland Institute, a Chicago-based conservative public policy think-tank, had organized the first international conference of climate change skepticism. The conference was titled *Global Warming Is Not a Crisis!*, and featured event elements common to any scientific event: seemingly esteemed keynote speakers, parallel sessions, and online proceedings.[5] The conference website stated that over 200 scientists from leading universities had participated in the event. For this controversy mapping exercise, I partnered with Andrei Mogoutov, the developer of a software tool for 'scientometric analysis' called ReseauLu, to examine the scientific publishing and citation networks of prominent speakers at this event.[6][7]

Our first query related to the apparent eventfulness of the inaugural Heartland conference. We wanted to know whether the scientific research and publication 'profiles' of climate skeptics were different from the profiles of non-skeptical climate scientists. More specifically, were the skeptics, beyond this specific conference, co-participants in a broader scientific community

4   Digital Methods Initiative, 'Issue Image Analysis', 2007, https://wiki.digitalmethods.net/Dmi/
     IssueImageAnalysis.
5   The Heartland Institute, 'First International Conference on Climate Change (ICCC-1)', 2008, http://
     climateconferences.heartland.org/iccc1/.
6   Scientometrics uses data sets of scientific publications and assesses these through citation analysis.
     More specifically, a scientometric analysis can extend from tracking citational behavior and referencing,
     to understanding these processes as constructing norms and rules of scientific writing, to considering
     how specific or groups of texts play out in an inter-referential network of influence and authority. P.
     Wouters, *The Citation Culture*, Amsterdam: University of Amsterdam, 1999.
7   See also A. Cambrosio, P. Cottereau, S. Popowycz, A. Mogoutov, and T. Vichnevskaia, 'Analysis of
     Heterogenous Networks: The ReseauLu Project', in B. Reber and C. Brossaud (eds.) *Digital Cognitive
     Technologies: Epistemology and the Knowledge Economy*, Hoboken, NJ: John Wiley & Sons, Inc, 2013.

dedicated to climate science? Or was it more accurate to understand them as a separate or differently networked or trained community (or on their way to becoming this), as the Heartland conference appeared to propose? In addition to this scientometric analysis, together with another summer school participant Bram Nijhof, I also *followed* the conference actors through to their personal websites to see whether these scientists wrote skeptical articles on topics *other* than climate change. This second research question is somewhat related to the first, and also straightforward: Should these actors best be considered as professional *climate science* experts that happened to be skeptical about specific findings or projections of climate change science data? Or were they skeptics in relationship to various controversies *as such* — writing critically or presenting as skeptics on a variety of subjects? Lastly, with Nijhof, I analyzed the hyperlinking behavior of these actors and their resonance within the top search engine results for the query of 'climate change'.[8] Upon discovering in these studies that the most prominent climate actors were skeptics first and foremost (as discussed in detail in Chapter 3), this geared me towards further studies of the controversy and its actors and ultimately led to the formulation of this book project.

Conducting National Analyses

In 2010, I was contacted by Denis Delbecq, a French climate journalist writing a dossier of several long-form articles about climate skepticism for the French environmental journal *Terra Eco*. Delbecq had come across my analysis of the Heartland actors on the mappingcontroversies.net platform and expressed interest in a similar collaboration with him that would apply these methods to an analysis of French climate science actors. He provided a list of prominent climate scientists (both climate skeptics and non-skeptics), including names of individuals and representative organizations. We used this data to conduct both hyperlink analysis (looking at the hyperlinks from the actors' websites) and resonance analysis (querying the prominence of these actors in the Google.fr search results for the query 'changement climatique'). Our results were published in *EcoTerra* and on Delbecq's blog, and resulted in the outing of a famous French skeptic, who had until then operated under a pseudonym.[9][10]

Soon after, in October 2011, the Royal Dutch Academy of Sciences (KNAW) published a report titled 'Climate Change: Science and Debate', aiming to articulate the current state of global climate science by delineating topics of consensus from those of controversy.[11] In response to these developments in the Netherlands, I collected a list of non-skeptical actors from the

8   These studies were published on the online research platform mappingcontroversies.net (as part of the EU 7th Framework project *Macospol*). S. Niederer, 'Climate Change Skeptics in Science', 2009, http://www.mappingcontroversies.net/Home/PlatformClimateChangeSkepticsScience.
9   D. Delbecq, 'A [F]rench Climate Skeptic Comes Out: He Is a Physicist', *Effets de Terre*, 2010, http://effetsdeterre.fr/2010/04/21/a-french-climate-skeptic-comes-out-he-is-a-physicist/. D. Delbecq, 'Dossier Climato-sceptiques', *TerraEco* (April 2010): 50–62.
10   D. Delbecq, and S. Niederer, 'Climatosceptiques et Climatologues, Quelle Place sur l'Internet?', 2010, http://effetsdeterre.fr/2010/04/12/climatosceptiques-quelle-place-sur-linternet/.
11   KNAW, *Klimaatverandering, Wetenschap en Debat*, Amsterdam: Koninklijke Nederlandse Academie van Wetenschappen, 2011, https://www.knaw.nl/nl/actueel/publicaties/klimaatverandering-wetenschap-en-debat/@@download/pdf_file/20101047.pdf.

contributors to the KNAW report, and a second list from the line-up of a skeptical gathering that was organized at Nieuwspoort in the Hague in critical response to the KNAW report, to conduct an analysis of Dutch climate skepticism similar to that of the French.[12] This made it possible to start to compare the two national situations. The Dutch study is discussed in detail in Chapter 3.

It was at this point that I found myself entering the controversy I was invested in researching, arguably in full accordance with Latour's directive that researchers jump straight into the middle of 'their' controversy as it unfolds. Following the publication of my work on these national climate change debates, Dutch actors, perhaps prompted by media monitoring tools of their own, started emailing me. In their messages to me, they included other scholars in cc (the 'carbon copy' setting in email. One email asked for a headshot to be placed alongside a review of my article. Another email described as 'hurtful' my linking of Dutch skeptics' work to research by Oreskes and others that discuss the financial ties of these actors to fossil fuels and other sponsoring industries. Others wrote to ask why I had not just contacted them personally to learn the truth about climate change, or posed my queries directly to them regarding their specific methodological approaches and tactics, assumedly to bypass the public nature and impact of my research findings. Somewhat taken aback by these direct responses (and also by their tone), I decided not to engage in direct conversation at that time.[13] Furthermore, observational distance is necessary for both of the approaches which I will introduce later in this chapter, namely 'content analysis' and 'digital methods', to keep their status as non-intrusive methods.

Formulating the Case Studies

As I further developed my research on the climate controversy on the web, I also sought the most suitable means to study a controversy of this nature that has no single communication channel but takes place across online platforms, resonating not only in mass media but also in search engine results, Wikipedia, Twitter and beyond. Important to note here is that these platforms have grown exponentially in the period of 2008 and 2015, the time during which I studied the debate, but that their status or role in controversies has never been systematically examined. Furthermore, during the same period, traditional mass media have had many struggles but have not disappeared. Rather, they have become part of, folded into, and entangled with the platforms and sources encountered when analyzing controversies through networked content. I considered that in order to understand specific controversies,

---

12   Nieuwspoort is a forum for political debate, situated next to the House of Representatives' building in the city center of The Hague. 'Nieuwspoort', http://www.nieuwspoort.nl/over-nieuwspoort/.

13   The question of how precisely I was able to label and split these actors as either skeptical or non-skeptical climate scientists I consider valid. Here, I followed the Latourian logic of there being no groups without 'group holders' and 'group talkers'. Bruno Latour, *Reassembling the Social*, Oxford: Oxford University Press, 2005. Somebody may not be a climate expert in daily life, but when this person is one of the editors of a publication on the climate controversy and consensus (in the KNAW example), they at that moment perform to identify with a 'group' of climate experts. Similarly, when opposing Dutch climate experts organize an event at Nieuwspoort to refute a scientific report as 'alarmist', they perform as skeptical 'group makers, group talkers, and group holders'. Latour, *Reassembling the Social*, 32.

as well as methods for the analysis of networked content through which they travel, media studies research would benefit from a deeper knowledge of the function or position that online platforms have in a controversy, and their entanglement with traditional mass media content. Hence, I decided to formulate case studies that could capture the climate change debates flowing through and across these online platforms.

To map and analyze the state and resonance of climate change actors and discourses through medium-specific digital methods, I included the use of websites through hyperlink analysis and search engine results, Wikipedia through interlinked articles and Twitter through its hashtags. Thus, my platform-specific case studies make use of different methodological approaches, taking the research outlook from controversy analysis and tools and methods developed in digital methods in order to further attune content analysis to networked digital media content. In the next section, I will address this research outlook provided by controversy analysis and very briefly discuss its roots in 'science and technology studies', before I formulate my main thesis and outline the case studies.

Traditions in Controversy Analysis

Controversy analysis, as previously mentioned, originated in science and technology studies (STS), and focuses especially on *scientific* controversies. Scientific controversies are said to 'destabilize' a system or convention of scientific truth claims, and in doing so reveal underlying dynamics of science and technology and their relations with a wider society that under normal circumstances tend to remain hidden.[14] STS scholars Trevor Pinch and Christine Leuenberger describe four influential approaches, which partly overlap chronologically, within STS-informed controversy analysis.[15] Firstly, the 'Priority Dispute studies' problematize claims towards who was the first scientist to make a particular scientific discovery. A second approach looks at the negative impacts — real or potential — of scientific and technological innovations (consider for example the political, social and ecological aspects of nuclear energy and genetic modification). A third key area of STS, as Pinch and Leuenberger note, is the Sociology of Scientific Knowledge (SSK), which emerged in the 1970s and operationalized the ideal of 'symmetry' to urge social researchers to 'use the same explanatory resources to explain both successful and unsuccessful knowledge claims'.[16] This principle can be applied especially well to scientific controversies, where different scientists each claim to present the truth and to refute the research methodology, argumentation, or outcomes of other(s). Symmetrical analysis enables the researchers of a controversy to study both (or all) sides of the story, including the scientific claims made by actors internal to the controversy object, by using 'the same sorts of sociological resources'.[17] Fourthly, Pinch and Leuenberger identify 'modern

---

14   T. Pinch and C. Leuenberger, 'Studying Scientific Controversy from the STS Perspective:
     Concluding Remarks on Panel "Citizen Participation and Science and Technology"',
     in *East Asian Science, Technology and Society*, 2006, http://fr.curriculumforge.org/
     TravaillongVincentr?action=AttachFile&do=get&target=Pinch+studying.pdf.
15   Pinch and Leuenberger, 'Studying Scientific Controversy from the STS Perspective', 4.
16   Pinch and Leuenberger, 'Studying Scientific Controversy from the STS Perspective', 12.
17   Pinch and Leuenberger, 'Studying Scientific Controversy from the STS Perspective', 12.

science and technology studies' that build heavily on SSK to regard controversies as 'integral to many features of scientific and technological practice and dissemination'.[18]

While STS has a strong tradition and methodological framework to study scientific controversies, it does not explicitly outline or champion specific digital methods for studying the digitally networked aspects of scientific knowledge communities. As the climate debate is not limited to offline media but also manifests itself across web platforms, there is a direct need for further methodological specificity. To analyze online networked content as part of a scientific (or other) controversy, we need to recognize the elaborate socio-technical formations — and transformations — of controversies in online networked content that impact the work and communities of scientific (and extra-scientific) truth-claims. Two of the schools of thought and practice I build my research techniques upon at this point, controversy analysis (as developed in education at Sciences Po, Paris) and 'issue mapping' (as developed by the Digital Methods Initiative at the University of Amsterdam) offer digital means of controversy analysis from similar scholarly traditions but with a distinct angle.[19] While the Parisian school stems from STS and operationalizes Actor-Network Theory to zoom in on a *controversy*, the Amsterdam approach builds on science and technology studies to track *issues* more broadly, be they controversial or not.[20][21][22][23][24]

This book makes integrative use of controversy analysis as well as digital methods (and tools) for issue mapping to conduct an analysis of the climate controversy across online platforms. As I outline in detail in the next chapter, a highly relevant research technique for both qualitative and quantitative analyses of mediated content precedes my work here, developed to study media content in the field of communication science under the name of 'content analysis'. Content analysis was incepted to study given or demarcated bodies of content (often referred to as 'texts' but not limited to that format), to analyze both formal features (e.g. the shot lengths of a television show, or the column widths and word counts of a printed text) and 'textual' meanings (broadly defined) including themes, tropes, recurring topics and terms, all in order

---

18   Pinch and Leuenberger, 'Studying Scientific Controversy from the STS Perspective', 5.
19   The third of which is content analysis, central to the next chapter.
20   N. Marres, 'Why Map Issues? On Controversy Analysis as a Digital Method', *Science, Technology & Human Values*, 0162243915574602, 2015, http://doi.org/10.1177/0162243915574602.
21   T. Venturini, 'Diving in Magma: How to Explore Controversies with Actor-network Theory', *Public Understanding of Science* 19.3 (2009): 258–273.
22   R. Rogers and N. Marres, 'Landscaping Climate Change: A Mapping Technique for Understanding Science and Technology Debates on the World Wide Web', *Public Understanding of Science* 9.2 (2000): 141–163.
23   Latour's *Mapping Controversies* educational program has culminated in the Médialab Sciences Po in Paris in 2009, which develops digital tools and methods for Controversy Mapping. Sciences Po's approach is 'interdisciplinary' and describes its work as 'seeking to apply computational techniques in order to detect, analyze and visualize public contestation over topical affairs'. Marres, 'Why Map Issues?'.
24   When analyzing controversy, researchers team up with programmers, data analysts, and information designers to create maps that make web content *differently* legible for further analysis. In my own research practice, I have worked in similar teams associated with the University of Amsterdam's Digital Methods Initiative, and participated in 'sprints' as part of the EU-projects MACOSPOL and EMAPS, in which we analyzed controversies through web data.

to make inferences about societal perceptions, cultural change, and trends in public opinion. A famous pre-web longitudinal content analysis study referenced in the scholarly literature is the *Cultural Indicators* program (of the 60s through 90s) by George Gerbner et al. that used weeklong aggregations of the prime-time television footage to record all representations of violence and construct 'violence profiles,' for this material. These representations were then interpreted and turned into 'cultural indicators,' which referred both to trends in network television's dramatic content and to viewer conceptions of social reality.[25][26] Content analysis has since been described as 'indigenous to communication research and [as] potentially one of the most important research techniques in the social sciences'.[27]

It is essential to emphasize that I understand content analysis to have always been inclusive of *potentially* all content types. By taking mass media as its most prominent raw data source, however, this kind of scholarship tended to be 'dominated by content analyses of newspapers, magazines, books, [radio] broadcasts, films, comics, and television programming' as one of its key scholars, Klaus Krippendorf pointed out.[28] Krippendorf, who I take to be centrally informative for my own work, has made explicit since content analysis' earliest methodological formation that (more or less publicly communicated) data of any kind could potentially be studied through content analysis. He mentions varieties of media 'content' as diverse as 'personal letters, children's talk, disarmament negotiations, witness accounts in courts, audiovisual records of therapeutic sessions, answers to open-ended interview questions, and computer conferences', and even 'postage stamps, motifs on ancient pottery, speech disturbances, the wear and tear of books, and dreams'. More theoretically, as a major proponent and methodological innovator of this field of media research, Krippendorff's assertion that 'anything that occurs in sufficient numbers and has reasonably stable meanings for a specific group of people may be subjected to content analysis', is a key driver of my own development of 'networked content analysis'.[29]

If, in practice, content analysis has mostly focused on neatly demarcated sets of texts or other media materials such as television shows, the specificity, dynamism, and networked nature of digital media content poses a myriad of new methodological challenges and opportunities to contemporary content analysts. Digital media content can be published or created on the World Wide Web, and enriched with opportunities for navigation and interaction. It can be networked by in-text hyperlinks (creating a so-called 'hypertext'), or by suggestions of related articles or other recommendation systems, or pulled into social media by prevalent 'Like' and 'Share' buttons on websites, urging users to link content to their own user profiles.[30] Online

25   G. Gerbner, 'Toward "Cultural Indicators": The Analysis of Mass Mediated Public Message Systems',
      Educational Technology Research and Development 17.2 (1969): 137–148.
26   G. Gerbner, 'Cultural Indicators: The Case of Violence in Television Drama', *The Annals of the American
      Academy of Political and Social Science* 388.1 (1970): 69–81.
27   K. Krippendorff, *Content Analysis: An Introduction to its Methodology*, first edition, Beverly Hills, CA:
      Sage Publications, 1980.
28   Krippendorff, *Content Analysis*, 404.
29   Krippendorff, *Content Analysis*.
30   C. Gerlitz and A. Helmond, 'The Like Economy: Social Buttons and the Data-intensive Web', *New Media
      & Society*, 2013, http://nms.sagepub.com/content/early/2013/02/03/1461444812472322.

content is *networked*. It is dynamic rather than stable; it often changes over time or moves from the front page to the archive. Social media further scatters content, offering a 'live feed' that is referred to as the qualitative and quantitative *real-timeness* of social media data, the content of which can be linked to, copied onto other networks, and archived across the (social) web.[31] These social media platforms each format, rank, and serve content in unique ways, which makes it important to start developing adaptive, digital methods that are attuned to the diverse specificities of these platforms.

Content analysis of such networked content may ask where the 'content' that is under analysis ends if all content is (more and less) meaningfully hyperlinked to other related content on other web pages. Indeed, how is it possible to demarcate a website? Is it methodologically appropriate to apply the techniques of content analysis that worked for printed newspapers like the *New York Times* or *The Guardian*, and for television formats such as *CNN* or *Al Jazeera*, to online news sites like www.nytimes.com and www.guardian.co.uk, let alone to a content search engine and aggregator like Google News? The answers to these questions as they have been offered by content analysis scholars throughout different phases in the history of the web are described extensively in Chapter 2, and can be summed up as broadly presenting two distinct approaches. The first, as described by McMillan, argues for standardization of methods towards the analysis of web content, which McMillan characterizes as a 'moving target'.[32] A second approach is formulated by Herring in response to McMillan, who proposes to combine traditional content analysis techniques with methodologies from disciplines such as linguistics and sociology to offer a more workable response to the challenges offered by 'new online media'.[33]

While these two approaches each offer ways forward for the analysis of web content, they are not concerned with the vast differences between different web platforms — the specific technicalities of which contribute significantly to the meaning of networked content. It is important to note that web content currently exists in and through the platforms and engines that produce it, which means a clean separation of content from its carrier is no longer feasible.[34] Different web platforms and search engines each carry their own (often visually undisclosed) formats and formatting; they have their own scenarios of use and their own terms of service; further, they also output their own results and rankings. Consider the example of Wikipedia, the collaboratively written encyclopedia project on a wiki, where each article has a page, sometimes other language versions, a discussion page, user statistics, a 'history' or archive of all previous versions of the article, all of which can be used in comparison with the current version of the article, as bots at work continue to edit text and undo vandalism.

---

31 L. Back, C. Lury, and R. Zimmer, 'Doing Real Time Research: Opportunities and Challenges', *National Centre for Research Methods (NRCM)*, *Methodological review paper*, 2012, http://eprints.ncrm. ac.uk/3157/1/real_time_research.pdf.

32 S. McMillan, 'The Microscope and the Moving Target: The Challenge of Applying Content Analysis to the World Wide Web', *Journalism and Mass Communication Quarterly* 77 (2000): 80–88.

33 S. Herring, 'Web Content Analysis: Expanding the Paradigm', in J. Hunsinger et al. (eds) *International Handbook of Internet Research*, Dordrecht: Springer, 2010, pp. 233-249.

34 Krippendorf stands out, as I emphasize in Chapter 2, in including this fact from the beginning, well before this research method had to deal with online networked content.

Differently for Twitter, the social network slash micro-blogging tool, user-broadcasted messages are bound by a limit of 140 characters per Tweet. They can include images, links to URLs, tags of other users (whether directly connected as 'followers' or not), and hashtags to network and aggregate individual content around specific events, issues, opinions, and themes. Content can include retweets of someone else's message (in several distinct ways, as described by Bruns and Burgess), which generates yet another layer to the networking of content.[35][36] These specificities of how platforms and engines serve, format, redistribute, and essentially co-produce content is what I refer to as the *technicity* of content.

Central Thesis: Accounting for Technicity

Controversy mapping, digital methods, and content analysis, in combination, offer means to study a controversy on the web that include this factor of technicity in the analysis of networked content. In this research, I will put forward such methods and techniques that take as their point of departure that the medium of the web now not only serves but also co-produces online content. The novel challenges posed by the dynamics of web content does not mean we have to dispose of content analysis altogether. On the contrary, as content analysis from the outset has been potentially inclusive of all varieties of content in and across contexts, its methods need to be amended only slightly — building on digital methods and controversy analysis — to suit the technicity of web content. I will argue that content analysis in its earliest form still offers model methods and approaches that, with appropriate amendments for the digital age, can be updated to stand as a strong methodological ground for what I name and develop here as 'networked content analysis'.

The central thesis of this study is that different web platforms and engines serve content with different technicities, which I argue are a crucial aspect of the object of study (i.e., web content) and should, therefore, be included in the analysis.[37][38][39][40] How can these insights

35   A. Bruns and J.E. Burgess, 'The use of Twitter Hashtags in the Formation of Ad Hoc Publics', in
     Proceedings of the 6th European Consortium for Political Research (ECPR) General Conference 2011,
     2011, http://eprints.qut.edu.au/46515.
36   A. Helmond, *The Web as Platform: Data Flows in Social Media*, Ph.D. Thesis, 19 June 2015, University
     of Amsterdam, Amsterdam.
37   R. Rogers, E. Weltevrede, S. Niederer, and E. Borra, 'National Web Studies: The case of Iran', in
     J. Hartley, J. Burgess and A. Bruns (eds) *Blackwell Companion to New Media Dynamics*, Oxford:
     Blackwell, 2013, pp. 142-166.
38   See also: R. König and M. Rasch, eds. *Society of the Query Reader: Reflections on Web Search*,
     Amsterdam: Institute of Network Cultures, 2014. What this research underlines is that the web may be
     'worldwide' in its infrastructure, but it is not in its access to content.
39   R. Deibert, J. Palfrey, R. Rohozinski, J. Zittrain, and M. Haraszti, *Access Controlled: The Shaping of
     Power, Rights, and Rule in Cyberspace*, Cambridge, MA: MIT Press, 2010.
40   Here it is important to point out that the attention to the technicity of content at the core of my
     research necessitates the recognition of the spatial organization and geo-location of content, as well
     as dislocation and censorship, which all problematize the very idea of a 'world wide web' of content
     assumed to be globally available. Internet censorship research has demonstrated how a user's geo-
     location is crucial to the availability of content, as served, for instance by the search engine Google.
     Research that critically comes to terms with these local differences in search engine results — which
     can be shown up by using a different language version of Google, or with VPN connections that access

from digital methods inform the application of content analysis to web content? As I am persistently emphasizing, developing means of collecting and analyzing digital media content across platforms starts with the problematic realization that each platform or engine has its own *technicity* and thus requires specific methods and analytical tools. To retain the strengths of content analysis for contemporary humanities and social research, and further develop techniques that better adapt to the specificities of networked content, the question central to this book is: how can technicity be meaningfully included in the analysis of online content?

In operationalizing this inclusive approach, I analyze the content of specific platforms alongside their technicity, for example, the user's access to read/write/link/archive capabilities, and identify the queries or tools that are necessary to demarcate and analyze content relevant to controversy objects that traverse these specific websites and platforms. Neither controversy analysis nor content analysis offers platform-specific techniques, which is why the addition of digital methods and tools is necessary for the analysis of such an interdisciplinary and popular, volatile public debate that is so widely distributed across platforms. In this way, I conduct what I consider to be useful, propositional forms, and methods of networked content analysis towards the study of the climate change debate online.

Networked Content Analysis of the Climate Debate

Climate change is defined by the United Nations Framework Convention on Climate Change (UNFCC) as the 'change of climate which is attributed directly or indirectly to human activity that alters the composition of the global atmosphere and which is in addition to natural climate variability observed over comparable time periods'.[41] The UNFCC distinguishes between human-attributed climate change and natural climate variability, a complex distinction that lies at the core of what is one of the most contentious and world-changing controversy objects of our time. There are clearly many reasons that I could propose for choosing to work with this complex issue in my development of networked content analysis methods. Quite apart from the political and scientific urgency accorded to this debate, as a new media researcher, I am particularly interested in the fact that to study climate change as a controversy object is to engage with a wide variety of (offline and online) media and knowledge spaces. Climate change remains on the agenda of NGOs and governments alike. Scholars have named it amongst the greatest threats (or 'risks,' to speak with Ulrich Beck) of our times and as a crisis of formidable scale.[42][43] This book does not contribute to climate *science* but instead focuses entirely on developing a networked content analysis of the climate controversy as

the web from other geo-locations — has been called 'search as research' by Rogers, and presented at international search engine research conferences such as the *Society of the Query*. R. Rogers, *Digital Methods*, Cambridge, MA.: MIT Press, 2013. R. Deibert, J. Palfrey, R. Rohozinski, J. Zittrain, and J.G. Stein, *Access Denied: The Practice and Policy of Global Internet Filtering*, Cambridge, MA: MIT Press, 2008.

41 United Nations, 'United Nations Framework Convention on Climate Change', 1992, https://unfccc.int/files/essential_background/background_publications_htmlpdf/application/pdf/conveng.pdf.

42 U. Beck, *World at Risk*, Cambridge: Polity Press, 2009.

43 B. Latour, 'Waiting for Gaia: Composing the Common World Through Arts and Politics', *Equilibri* 16.3 (2012): 515–538.

it is specifically mediated and transformed by online platforms and actors, in order to gain insight in how such controversial debates evolve and how certain actors and viewpoints may resonate more forcefully than others. Accordingly, the next section will introduce prior studies in climate-related content analysis by Anthony Downs, building beyond the work that opened this introduction.

Before reappraising Downs, it is necessary to specify further my research outlook. Where my central concern here is to develop means to include technicity in the analysis of networked content, I am dealing with the specificity of the question by applying it to the topic of web content on climate change. Looking at how technicity can be included in the analysis of networked climate change content, I take to three online platforms that each represent a different web culture, if you will. The web as accessed through the search engine Google is for many Internet users the main point of access to web content.[44] Twitter is one of the most prominent social platforms online, with its content available through an API. Wikipedia is the most-used online equivalent of an encyclopedia. As climate change is present across distinct sites of knowledge sharing, discussion and dissemination (science, news, popular media) it can be studied across platforms and analyzed in terms of: the variety and prominence of actors and sources (Google); the online dynamics of knowledge production (Wikipedia); and the sub-issues of climate change as shared online (Twitter).

Building upon the strengths of existing content analysis projects, my formulation of networked content analysis asks what may be learned from previous applications of content analysis. How has content analysis been amended since its very first application to web-based content? In applying networked content analysis to online climate change content, I will address how the issue of climate change can be studied there (via Google/Wikipedia/Twitter) and identify the specific technicities of such content. Given that the study of climate change across media has already been strongly attended to in earlier content analysis studies, I briefly discuss this research pre-history and its relevance to my own work in the next section.

The Climate Change Debate as an Object of Study

Climate change as an issue has, in fact, been attended to with fine-grained content analysis methods since the early seventies. In his article *Up and Down with Ecology: The Issue-attention Cycle,* Anthony Downs described how the environment, like any societal issue, is subject to a rise and fall in public interest. He uses the notion of the 'issue-attention cycle' to describe common dynamics in public attention that occur for 'most key domestic issues'.[45] Downs' articulation of the issue attention cycle knows five stages: (1) the pre-problem stage, (2) alarmed discovery and euphoric enthusiasm, (3) realization of the cost of significant prog-

44   The dominance of Google Web Search has been critically assessed by scholars including Carr, Lovink, and Vaidhyanathan. See: N. Carr, *The Big Switch: Rewiring the World, from Edison to Google,* New York, NY: W.W. Norton & Company, 2008. G. Lovink, 'The Society of the Query and the Googlisation of Our Lives: A Tribute to Joseph Weizenbaum', *Eurozine,* 2008, http://www.eurozine.com/articles/2008-09-05-lovink-en.html. S. Vaidhyanathan, *The Googlization of Everything: (And Why We Should Worry),* Berkeley, CA: University of California Press, 2011.
45   A. Downs, 'Up and Down with Ecology: The Issue-attention Cycle', *The Public Interest* 28 (1972): 38.

ress, (4) gradual decline of intense public interest and, lastly, (5) the post-problem stage.[46] Downs sees the 'remarkably widespread upsurge of interest in the quality of our environment' as involving such an issue-attention cycle, in which the 'change in public attitudes has been much faster than any changes in the environment itself'.[47] Downs' work has been subjected to strong criticism, mainly on the linearity assumed by his proposed cycle model, and on the research's focus more on mediation as such, over the mediation of this specific and urgent issue, as described thoroughly by McComas and Shanahan.[48] With these qualifications, analysts of media content have taken up Downs' approach and further extended its application to environment-related issues.

In what they refer to as a '(de)construction' of the issue-attention cycle for environmental issues, McComas and Shanahan compare the climate change news coverage of the major US newspapers, *The New York Times* and *The Washington Post*, between 1980-1995.[49] Their research confirms the cyclical nature of attention to the issue of climate change, and even recognizes different stages that dialogue with Downs' own, in which:

> [T]he implied danger and consequences of global warming gain more prominence on the upswing of newspaper attention, whereas controversy among scientists receives greater attention in the maintenance phase. The economics of dealing with global warming also receive more considerable attention during the maintenance phase and downside of the attention cycle.[50]

Where these researchers stress the importance of the 'role played by narratives in driving media attention to environmental issues', others have stressed how real-life events (such as extreme weather) are a crucial catalyst in the garnering of public attention for an issue of 'celebrity status'.[51][52] A concept that builds on this analytical approach to issue-attention is the 'news spiral', which refers to the phenomenon that once the climate is in the news, this creates a general upsurge of interest in (and reporting on) other environmental issues.[53] The retrieval and analyses of attention and news cycles fit into the ongoing methods and applications of content analysis at large.

---

46   Downs, 'Up and Down with Ecology', 39-40.
47   Downs, 'Up and Down with Ecology', 38.
48   K. McComas and J. Shanahan, 'Telling Stories About Global Climate Change Measuring the Impact of Narratives on Issue Cycles', *Communication Research* 26.1 (1999): 30–57.
49   McComas and Shanahan, 'Telling Stories About Global Climate Change Measuring the Impact of Narratives on Issue Cycles'.
50   McComas and Shanahan, 'Telling Stories About Global Climate Change Measuring the Impact of Narratives on Issue Cycles', 30.
51   McComas and Shanahan, 'Telling Stories About Global Climate Change Measuring the Impact of Narratives on Issue Cycles', 33.
52   S. Ungar, 'The Rise and (Relative) Decline of Global Warming as a Social Problem', *The Sociological Quarterly* 33.4 (1992): 483–501.
53   M. Djerf-Pierre, 'When Attention Drives Attention: Issue Dynamics in Environmental News Reporting Over Five Decades', *European Journal of Communication*, 27.3 (2012): 291–304.

Chapter 2 discusses the early disciplinary formation of content analysis and develops an approach towards networked content analysis. Content analysis has a strong history of use in communication science, where large bodies of text are analyzed for features or (recurring) themes, in order to identify cultural indicators or make other inferences about the text. To apply these methods to web content remains a challenging exercise to researchers of various scholarly disciplines, for, unlike traditional print media such as newspapers or books, web content is often dynamic. It is also networked, which poses problems for the demarcation of the content under study. To grapple with these technical specificities of web content, research-ers either stay close to traditional content analysis techniques or choose to pull in methods from other disciplines and seek more extended paradigms of web content analysis.[54] [55] [56] In this chapter, I will give an overview of these strategies preceding my research, and introduce novel means of networked content analysis that include the technicity of web content as part of the analysis and repositions content analysis (in the tradition of Krippendorff) as a medium-specific approach.

In the three case studies that follow this methodological discussion, I assess the climate change debate on different platforms. As the climate debate does not only take place across platforms, but also over time, the studies presented will assess diverse moments in the climate change debate, ranging from the first skeptical conference of 2008 to the 'Paris Agreement' of 2015. The aim of the study is not to present a neat chronology or timeline of the debate from beginning to end, nor, at the other extreme, to do away with historical analyses. The point of entry is less the debate's transformation over time (or its timing), than its entangled relation to the platform and its specificities. How can we amend content analysis to attune to the technicity of networked content, knowing for instance that on the web, search engines rank content, websites are hyperlinked, and actors in one issue may also be working on another issues and publishing about this on their personal websites? And what does the platform, or the engine, do to the debate?

In Chapter 3, I analyze the networks of climate debate actors using search engines and scientometric analysis. This chapter uses search engines (ISI and Google) and hyperlink analysis to research the place and status of climate skepticism within both climate science itself and the climate change controversy as it takes place beyond the scientific literature. Here, the central question is how networked content analysis can be conducted *through* the web, taking into account the technicity of search engines. The case study zooms in on climate change actors and their prominence, as identified by search engines. It asks how the technical logic of search might be used to measure and compare the prominence of actors in a specific issue, in this case, by looking at the resonance of climate change scientists (both skeptical and non-skeptical) within a demarcated set of websites. Hyperlink analysis and search engines enable comprehension of the group formation of actors in the debate and measure their resonance within web sources on the issue of climate change. Traditional sci-

---

54   McMillan, 'The Microscope and the Moving Target'.
55   Herring, 'Web Content Analysis'.
56   C. Weare and W. Lin, 'Content Analysis of the World Wide Web: Opportunities and Challenges', *Social Science Computer Review* 18 (2010): 272–292.

entometrics paired with digital methods offer a detailed picture of the status, group formations and issue commitments of climate change skeptics, and questions whether their interest lies in skepticism itself or in climate change.

In Chapter 4, I discuss Wikipedia as a socio-technical utility for climate controversy mapping. The technicity of Wikipedia content makes it possible to refine further the techniques of networked content analysis, so as to enable matters of resonance, relational dimensions of content, actor engagement and controversy management to be studied within this encyclopedia project. Wikipedia, as a wiki-based encyclopedia platform, offers various levels of access to information on article histories and editors, enabling researchers to 'follow the actors' and close-read their positions, references and commitment to a specific issue. In this case study, I discuss how Wikipedia has been researched since its launch in 2001, and how dominant research practices have disregarded some of the crucial technical specificities of Wikipedia and the production, organization and maintenance of its content. I then zoom in on more recent controversy analyses, attending to the technicity of Wikipedia content by looking at discussions on the talk pages (for the article on Gdańsk/Danzig), and by conducting a comparative analysis of articles across language versions (for the case of the Srebrenica massacre). Lastly, I discuss a networked content analysis of climate change related articles, tracing its networked content and close reading its actor behavior. I discuss a climate change article ecology study from 2009 and the development of a Wikipedia controversy analysis tool developed in 2014 titled 'Contropedia'. I propose here to treat Wikipedia as a data-rich site of social research, through a networked content analysis of climate change articles and their linkages.

In the final case study of Chapter 5, I conduct a networked content analysis of climate change-related Twitter messages (or 'tweets') to map the state of the climate change debate online. Here, I analyze Twitter data to consider four related climate change discourses: adaptation (to climate change), skepticism (towards the man-made origins and unprecedentedness of climate change), mitigation (the prevention of further climate change by minimizing its causes), and conflict (here taken to mean political unrest relatable to so-called 'climate change vulnerability').[57] Given climate vulnerability has recently become a prominent and focalizing discourse within climate change, both in the scientific literature (as mapped out by the IPCC in 2014) and in news coverage around climate change, I will zoom in on this issue in more detail. Recently, new debates concerning climate change research and modeling have arisen as experts are increasingly drawing connections between climate vulnerability and human conflict. Major news media outlets increasingly contribute to circulating an understanding of climate change vulnerability as a potential factor in social unrest, including in Syria and Egypt, explaining how drought and water scarcity may have intensified the Arab Spring. Twitter already has a strong tradition of being repurposed to study events, uprisings and social responsiveness to the news.

---

57   In the EMAPS Digital Methods Fall Data Sprint of October 2013, we asked whether conflict could be seen as a fourth phase in the evolution of the issue of climate change, after skepticism, mitigation and adaptation. EMAPS, 'Vulnerability, Resilience and Conflict: Mapping Climate Change, Reading Cli-fi', *Electronic Maps to Assist Public Science Blog*, 2013, http://www.emapsproject.com/blog/archives/2293.

In this chapter, to study Twitter's climate content and include its technicity, I create keyword profiles and additionally zoom in on the hashtags used within a set of climate change tweets. A co-hashtag analysis of this set of tweets reveals an ecology of climate change-related sub-issues illustrating the current state of climate action and adaptation — a multifarious presence of vulnerability variables related to data sets on animals, habitats and more, affected by extreme weather conditions. In attending to a descriptive analysis of sub-issues within the climate change controversy, which has such complex social dimensions, this chapter exemplifies how controversy does not end once consensus on some aspects of the science is publicly secured.

Chapter 6 holds the conclusions, in which I discuss the findings of the various case studies on two levels: that of the methodological toolbox of networked content analysis as well as on the level of the controversy mapping itself, reiterating what the various case studies teach us about the climate change debate, and gather up implications for the practice of networked content analysis. Taking the lessons learned from the case studies on the study of the climate debate with Twitter, Wikipedia, and Google, I return to Krippendorff to revisit his foundational work and propose appropriately amended techniques and tools for networked content analysis. Subsequently, I discuss the challenges for future research. As the climate controversy plays out on many platforms that, in turn, *pull in* traditional mass media content, I show how combined and interlinked findings across platforms provide a more comprehensive mapping of a multi-platformed issue.

2. FOUNDATIONS OF CONTENT ANALYSIS

The drastically changing nature of content in the move from print and elsewhere (e.g.,television) to the web has challenged the techniques and tools of content analysis, which, upon its inception, concerned itself mostly with large but static groupings of texts. Unlike modern print media such as newspapers or books, web content is often unstable and dynamic. It is also networked, which poses more problems for the researcher regarding the demarcation of the 'text' under study. Before further exploring this difference that technicity makes when aiming to do content analysis across the web, it is necessary to review the foundational status, methodologies, and tools of content analysis that existed as developed for (pre-web) mass media content. This chapter offers a historical perspective on the foundations of content analysis, discussing its scholarly roots and exploring how it has been modified as a field of research along with the changing technicities of content that it engages with. My historical reappraisal of the concepts and methods of content analysis considers first the work of Klaus Krippendorff, a groundbreaking content analysis scholar and, not coincidentally, a co-organizer of the first content analysis conference at Annenberg in 1969. After a brief examination of the foundational work by Berelson and Gerbner, I will come to describe Krippendorff's seminal work *Content Analysis: An Introduction to its Methodology,* in which he lays out the requirements of a sound content analysis research framework.[1]

Secondly, I will address the challenges this approach faces since the computer has become more of a content producer and site of production and publication, rather than merely a research aid for large-scale analyses of 'texts', broadly defined. Here, I will build on responses to the work of Krippendorff by communications and advertising scholar Sally McMillan and linguist and information scientist Susan Herring, who further developed Krippendorff's techniques to grapple with the technical specificities of web content, which I refer to as its technicity. The term technicity, as described in the introduction, refers to the technologically composed nature of web content — the fact that content can hardly be separated from its carrier (a specific web platform for instance), and that technical agents such as hyperlinks and shares are not mere features, but *part of* the content under study.[2][3] Accordingly, when looking at previous applications of content analysis to web content, I ask how the *technicity* may be made part of the definition, collection, and analysis of content being studied, which is the central question of this book.

---

1   In this chapter, I will refer mostly to the second edition published in 2004, as this was thoroughly revised to describe the analysis of 'computer readable' content and presents a more mature method and technique of content analysis since the first edition of 1980. K. Krippendorff, *Content Analysis: An Introduction to its Methodology*, second edition, Thousand Oaks: Sage Publications, 2004, xiv. I will occasionally refer to the third edition of 2013, e.g., when addressing recent discussions or techniques not included in the previous editions. K. Krippendorff, *Content Analysis: An Introduction to its Methodology*, third edition, Thousand Oaks, CA: Sage Publications, 2013.

2   S. Niederer and J. van Dijck, 'Wisdom of the Crowd or Technicity of Content? Wikipedia as a Sociotechnical System', *New Media & Society* 12.8 (2010): 1368–1387.

3   S. Niederer and J. van Dijck, 'Wisdom of the Crowd or Technicity of Content? Wikipedia as a Sociotechnical System', in M. David and P. Milward (eds) *Researching Society Online*, London: Sage, 2014.

Thirdly, I will ask whether content analysis should be enhanced to suit the analysis of networked and dynamic information online. Looking at the traditions in content analysis, a return to its roots may prove more productive. I would argue that conventional content analysis still holds valuable insights for current (online) approaches of web content. However, what needs to be explored are the necessary steps towards networked content analysis that takes the technicity of web content and the variety thereof as its point of departure. Lastly, I will describe how I will apply networked content analysis to study the issue of climate change, in the case studies in this book. I underline the importance of the issue for our day and age, but also describe strong preceding research in the content analysis of climate change content.

Emergence of a Research Field

The field of content analysis considers its first seminal work to be that of Berend Berelson of 1952, titled *Content Analysis in Communication Research*, which describes content analysis as an important research technique for social scientists and media scholars for reading social and cultural change from (the analysis of) mediated messages.[4] For example, in a study from 1948, Berelson and Salter study prejudice against minority groups through the analysis of popular magazine fiction.[5] In the same tradition, as mentioned in the Introduction, George Gerbner has studied violence on television and the representation of for instance women and children during primetime programming, to derive *cultural indicators,* the indicators of their position in society at a given time.[6]

Scholars often refer to the inclusion of the definition of 'content analysis' in Webster's Dictionary of the English Language in 1961 as a milestone in the establishment and public recognition of the field. Here, content analysis was defined as the 'analysis of the manifest and latent content of a body of communicated material (as a book or film) through classification, tabulation, and evaluation of its key symbols and themes in order to ascertain its meaning and probable effect'.[7] In November of 1969, another milestone took place with the content analysis conference of the Annenberg School of Communications, where over 400 scholars gathered from approximately 85 educational and scientific institutions in the United States and Canada to discuss the application of content analysis to and from a wide range of academic disciplines.[8][9] The conference also featured a panel dedicated to 'Computer Techniques in Content Analysis and Computational Linguistics', focusing solely on different ways in which content could be analyzed by the computer and by computer-aided content analysts. The scholars who presented computational analyses, in particular, at this inaugural event also

4   B. Berelson, 'Content Analysis in Communication Research', 1952, http://psycnet.apa.org/psycinfo/1953-07730-000.
5   B. Berelson and P.J. Salter, 'Majority and Minority Americans: An Analysis of Magazine Fiction', *The Public Opinion Quarterly,* 10 (1948): 168–190.
6   Annenberg School for Communication, *George Gerbner Archive,* University of Pennsylvania, 2006.
7   A. Merriam-Webster, *Webster's New Collegiate Dictionary,* G.&C. Merriam Company, Publishers, 1961.
8   Presently called Annenberg School for Communication. 'Annenberg School for Communication', https://www.asc.upenn.edu/.
9   G. Gerbner, O. Holsti, K. Krippendorff, W.J. Paisley, and P.J. Stone, eds. *The Analysis of Communication Contents: Development in Scientific Theories and Computer Techniques,* Wiley, 1969, xiii.

came from a diverse set of fields, including 'political science, psychiatry, sociology, English, and social psychology'.[10][11] It is worth keeping these early, partially interdisciplinary beginnings in mind when negotiating contemporary applications of content analysis by different academic fields. With the more recent infusion of culture with information technology, content analysis' early trajectory, as well as its focus on text and image analysis, merges with the interests of information science and allied fields in data-driven contemporary cultural analysis; this situation and convergence of practices and methods continues to create confusion about the possibilities of techniques for studying culture through content.

The most significant disciplinary figure of early content analysis, Klaus Krippendorff, defines content analysis as a 'scientific tool' and 'a research technique for making replicable and valid inferences from text to the contexts of their use'.[12] He deployed terms and concepts from outside the qualitative humanities normally concerned with content, like for example 'scientific', 'replicable' and 'valid', to emphasize the need for formalization of techniques and tools of analysis. At the same time, however, his use of the word *text* does not refer only to written materials but expansively may include 'works of art, images, maps, sounds, signs, symbols, and even numerical records' and other data.[13] Krippendorff makes the significant conceptual point that it is precisely one's definition of what content is, and how that is delimited, that leads to specific kinds of analytical results. As we will see with the analysis of networked content, it is indeed this refinement of definitions and approaches to the time and materiality of 'content' that needs to be amended. This is important for the recognition of the technicity as an active material agent and *part of* the content, rather than as a challenge that disturbs or supposedly renders difficult the demarcation and study of content online.

In other words, how one chooses to define content paves the way for specific research questions, methodological choices, and analytical consequences to play out over others. Content analysis, in this sense is not an entirely standardized or standardizable practice but is applied *across* scholarly disciplines that have used many different strategies of coping with the challenges posed by content on the web. Krippendorff dates this broadening as coinciding with some of the earliest applications of content analysis to the (further) growth of mass media after WWII. This rise of the field of content analysis to deal with expanded media formats, he argues, meant a loss of focus already then, as 'everything seemed to be content analyzable and every analysis of symbolic phenomena became Content Analysis'.[14] Krippendorff describes how various disciplines began to apply the research techniques of content analysis differently:

---

10 Stone in Gerbner et al. *The Analysis of Communication Contents,* 335.
11 It is worth mentioning here that at this historical moment, the computer being brought to work on content analysis was, specifically, a machine reading text from punch cards or microfilm, or otherwise dealing with content 'typed in from a computer console'. Accordingly, the approaches to content analysis presented were often captured in pieces of software and developed in different ways that directly reflected the specific state of the technology. Some approaches were programmed by the scholars themselves or programmed by others, including technicians, under close supervision from scholars, while yet other scholars completely outsourced programming responsibilities in full. Stone in Gerbner et al. *The Analysis of Communication Contents,* 336.
12 Krippendorff, *Content Analysis,* 2004, 24.
13 Krippendorff, *Content Analysis,* 2013, 25.
14 Krippendorff, *Content Analysis,* 2004, 12.

ethnographers were interacting with their informants (something content analysts usually do not do, as they prioritize 'unobtrusive' analyses) and also analyzing their own personal field notes as 'content', while social scientists studied educational materials to identify societal trends. At this point, Krippendorff develops a conceptual framework for content analysis that serves not only as a tool with which to (re-)establish a focus for this research methodology but also as a practical, analytical and methodological guide for researchers to *apply* the methods to diverse types of content. In the next section, I will describe this framework as introduced by Krippendorff and briefly reflect on how its components may hold in networked.

Krippendorff's framework lays out six components necessary for a content analysis research project, all of which are to be included though not necessarily in this sequential order:

- A body of text, the data that a content analyst has available to begin an analytical effort;
- A research question that the analyst seeks to answer by examining the body of text;
- A context of the analyst's choice within which to make sense of the body of text;
- An analytical construct that operationalizes what the analyst knows about the co text;
- Inferences that are intended to answer the research question, which constitute the basic accomplishment of the content analysis;
- Validating evidence, which the ultimate justification of the content analysis.[15]

Importantly, from the beginning point of his procedural outline, Krippendorff does not describe how content should be collected for well-formed content analysis to take place. The content to be analyzed seems not in question, in the sense that the text is already assumed to be accessible to the scholar (as, for example, a set of recent newspaper articles might be), demarcated, and readily available for study. The formulation of the research question and context narrows the broad scope of content analysis' disciplinarity slightly more. (Again, the term 'text' also refers to images, websites, music, etc.) In the next outlined step, Krippendorff emphasizes the importance of tailoring appropriate research questions, when stating that — in contrast to a deliberately open-ended interpretive approach to texts — strong research questions enable the researcher to read a text with more analytical distance. This allows the analyst not to just follow the author (in the Latourian sense described in the introduction) in what the actor says is in the text but instead, read off content with a specific question in mind. In this sense, the research question could also be described as a methodological tool in itself, with which to create a selection or sample of the data appropriate for answering the question.

As Krippendorff asserts, '[data] become[s] text to the analyst within the context that the analyst has chosen to read [it], that is, from within the analysis'.[16] The analyst's background and scope and the research questions in combination provide the texts with a novel interpretive mechanism, within which they can be analyzed. A political scientist and an anthropologist might analyze the same piece of text differently, for instance. With regards to the analytical

---

15   Krippendorff, *Content Analysis*, 2004, 29-30.
16   Krippendorff, *Content Analysis*, 2004, 33.

construct, Krippendorff stresses the importance of the *research* context in which a given text 'would arguably make sense'.[17] The analysis of text should be conducted in line with what is known about its uses. Krippendorff's fifth point constitutes the core of content analysis, in so far as the analysis enables the researcher to make inferences that scale appropriately. Krippendorff emphasizes the strength of abductive inferences — meaning those findings that are made across 'logically distinct domains' where multiple variables are taken into account — and compares this approach to the logic of reasoning employed by Sherlock Holmes, who uses clues to solve or sort through a larger reality and situation.[18] For example, one can date a text by analyzing the vocabulary it uses, or infer the poignant issues of a city by studying letters sent to the municipality or local newspaper.[19]

Lastly, and clearly in the interest of *not* letting abductive inferences over-reach, or otherwise become scientifically suspect, Krippendorff argues that all content analyses should be 'validatable in principle'.[20] Importantly for Krippendorff, this means there is a necessity to enable correlation of the research results with other data or information that stands *outside* the scope of the original analysis. The question of when the data requires a baseline outside of the content under study is one that resonates in the study of web content.[21] In the realm of content analysis, this discussion has also taken place, including the suggestion of validating mass media content analysis (of culture) with audience interviews.[22] For example, Gerbner on multiple occasions tried to correlate his Cultural Indicators research on violence in prime-time television with a survey on whether people also concurrently perceived the world as a violent place.[23]

While the definition and demarcation of content were never that straightforward in the case of offline mass media materials, the rise of digital media has further complicated these matters. Digitization of content changed the nature of the materials already, raising new questions (e.g., Should column-width still be considered?). With hyperlinks, content became networked and thus harder to demarcate (Where does this content end?). Search engines brought about new ways of presenting and ranking data (What is the most important source?), and platformization gives shape to the far-stretching entanglement of social media with other web content.[24]

As I will discuss in this chapter, the defining characteristics of web content pose new challenges to the above outlines, conditions, and expected consequences of what once fell under the purview of content analysis. To make a move into what I name networked content analysis, namely the application of content analysis on the web and the challenges thereof, it is

---

17   Krippendorff, *Content Analysis,* 2004, 35.
18   Krippendorff, *Content Analysis,* 2004, 38.
19   Krippendorff, *Content Analysis,* 2004, 42.
20   Krippendorff, *Content Analysis,* 2004, 39.
21   See for instance R. Rogers, F. Janssen, M. Stevenson, and E. Weltevrede, 'Mapping Democracy', in *Global Informaton Society Watch*, The Hague: Hivos, 2009, pp. 47-57.
22   Krippendorff, *Content Analysis,* 2013, 44.
23   G. Gerbner, L. Gross, N. Signorielli, M. Morgan, and M. Jackson-Beeck, 'The Demonstration of Power: Violence Profile No. 10', *Journal of Communication* 29.3 (1979): 177–196.
24   See also Helmond, *The Web as Platform.*

important to engage with the challenges of this transition as these have been pre-conceived and processed by scholars identifying with the foundations of the field. This includes the work of Sally McMillan, who describes the study of web content as like looking at 'a moving target through a microscope'.[25] Web content in the late 1990s was in many respects different from web content in 2009 or 2014; this is a fact that should never be lost hold of. In the late 1990s, which is the period in which the studies McMillan reviews in her paper were conducted, the web did not yet have 'platforms' and was still in its early days of search engines and web browsers. Content was, however, already networked by hyperlinks and website *features*, which thus were the focus of many analyses of this period.

Web Content Analysis: A Moving Target Seen Through a Microscope (McMillan)

In her article *Web Content Analysis: A Moving Target Seen Through a Microscope*, notably included in Krippendorff's 2004 edited volume *Content Analysis: An Introduction to its Methodology*, McMillan takes stock of the challenges researchers face when applying traditional content analysis techniques to the study of web content. Interestingly, McMillan takes up certain directives from Krippendorff's original content analysis framework to systematically track the present theoretical varieties of contemporary content analysis methods and theories in this paper. Firstly, McMillan compiles a collection of papers by searching the Social Sciences Citation Index (SSCI) for the keyword combinations 'web' and 'content analysis' as well as 'internet' and 'content analysis'. Secondly, McMillan seeks papers from selected communication journals as well as communication conferences not indexed in the SSCI. Finally, she expands the list by checking the bibliographies of all the found sources, and adding relevant cited studies to the list. In all, she finds a total of nineteen studies dedicated to content analysis on the web and another eleven studies that are dedicated to the analysis of other digital content, such as email and ListServs, both of which were important online media at the time but which are not included in her final study.[26]

Having collected her sources, McMillan relies on a research protocol close to Krippendorff's, checking each study for the resemblance of its components and methods to the original content analysis framework.[27][28] She then compares the 19 articles to identify how the challenges of applying content analysis to web content research were being dealt with by each of the authors. From this, McMillan induced five steps that in her view, should be part of all web content analysis studies, and which should be compared to Krippendorff's original list above:

1. Formulate the research questions and/or hypotheses;
2. Create a sample;

---

25   McMillan, 'The Microscope and the Moving Target', 80.
26   McMillan, 'The Microscope and the Moving Target', 88.
27   McMillan, 'The Microscope and the Moving Target'.
28   Krippendorff, *Content Analysis*, 2004.

3. Further define categories:
  (a) Establish the time period of the study, as web research calls for rapid collection of data;
  (b) Identify context units;
  (c) Develop coding units;
4. Train the coders and check the reliability of their coding skills;
5. Analyze and interpret data.

I want to discuss this paper in more detail because McMillan does try to address the issue of content collection that goes unstated in Krippendorff. Firstly, aiming to summarize how scholars collected their content, McMillan carefully lists the sampling strategies she has found in her list of 19 studies. She notes a wide variety of ways in which the researchers compiled their collections of websites to be analyzed. Most of the studies identify existing lists of reputable sources. In a footnote, McMillan issues a warning for web researchers using search engines, a novel tool at the time, stressing the importance of knowing as much as possible about how a search engine chooses and prioritizes its results before deciding how to use it for sampling. (I will more fully elaborate on this issue through the case study and argumentation of chapter 4 that deals with search engine results for the analysis of the position and resonance of climate skepticism on the web.)

Comparing McMillan's assembled lists of steps to the original provided by Krippendorff, one important component is now missing, which is validatability. This omission, I argue, very directly points to one of the key problems in using traditional content analysis methods without alteration in the analysis of web content: the fact that validation, which presumes an offline reference as a baseline, is not always possible in the analysis of digital and networked content. I would propose that an offline validation of online research in many cases is impossible. Thought-provokingly, scholars have asked in which cases the online *is* the only relevant baseline.[29][30] Linguist and information scientist Susan Herring recognizes that web content is indeed a different kind of object compared to the pre-web content of Krippendorff's time. In her 2010 paper, *Web Content Analysis: Expanding the Paradigm*, she calls for a widening of the research paradigms and methods attendant to web-oriented content analysis.

Web Content Analysis: Expanding the Paradigm (Herring)

Rather than proposing wholly novel means of analysis, Herring proposes a combination of methods from various disciplines that can help the analyst to research the new kinds of content that occur on the web. Herring begins her contribution by noting the semantic differentiation between 'web [content analysis]', a narrower kind of research where traditional content analysis methods are applied to the web, and '[web content] analysis', or what she calls WebCA, which is the analysis of web content in a broader sense, where various 'traditional

---

29 D. Moats, 'From Digital Methods to Digital Ontologies: Bruno Latour and Richard Rogers at CSISP', 2012, http://www.csisponline.net/2012/03/12/from-digital-methods-to-digital-ontologies-bruno-latour-and-richard-rogers-at-csisp/.

30 R. Rogers, *The End of the Virtual: Digital Methods*, Amsterdam: Vossiuspers UvA, 2009.

and non-traditional techniques' can be applied.[31] Herring promotes the latter by showing how traditional content analysis can be combined with methodologies from disciplines such as linguistics and sociology to offer a more workable response to the challenges offered by 'new online media'.[32] She illustrates this with examples from blog analyses and conversations online.

Herring's 'more general' definition of web content covers 'various types of information "contained" in new media documents [...], all of which can communicate meaning'.[33] This definition is very similar to the earlier definitions of content analysis that were critiqued by Krippendorff, for the presumption that content is 'contained in messages, waiting to be separated from its form and described', as the true nature of content 'resides *inside* a text'.[34] The broadening that Herring proposes is, in fact, a return to another specific idea of content, where various content types (all of which Krippendorff would refer to as *text*) can each communicate meaning. The broadening of the paradigm in her paper's title refers on the one hand to the inclusion of the analysis of these various types of online content. In other words, besides the more traditional content *elements* that might be considered by content analysts, such as images, themes, and features, she includes a range of newer online-only (or: natively digital) elements, such as the hyperlink.

Furthermore, Herring argues that the research practice she denotes as '[web content] analysis' would benefit from a broadening of its methodology, by including methods from other disciplines (see Figure 1). From sociology (and social network analysis), it is possible to attend to link analyses, from communication science (and content analysis) one can do feature analyses, and from linguistics (and discourse analysis) the contributing methodologies make it possible to produce computer-mediated discourse analysis. Rather than proposing medium-specific approaches to 'web content analysis', she proposes to broaden the methodological apparatus, by including other non-web-specific methods from different disciplines.

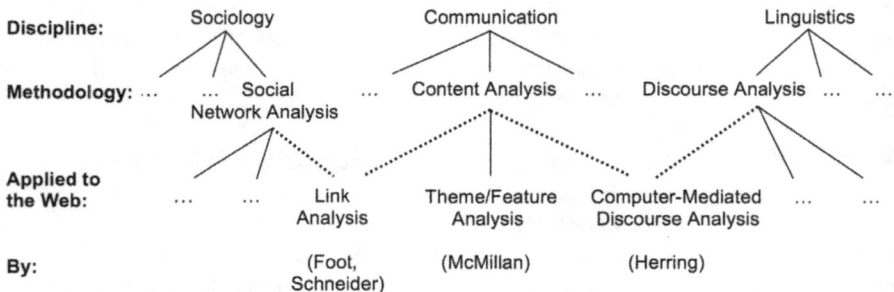

Discipline:	Sociology	Communication	Linguistics
Methodology: ...	... Social Network Analysis	... Content Analysis ...	Discourse Analysis ... ...
Applied to the Web:	... ... Link Analysis	Theme/Feature Analysis	Computer-Mediated Discourse Analysis ... ...
By:	(Foot, Schneider)	(McMillan)	(Herring)

Figure 1: Widening of the content analysis paradigm. Herring's brief overview of approaches to web content analysis, as modelled in Web Content Analysis *(p.240).*

31   Herring, 'Web Content Analysis', 235.
32   Herring, 'Web Content Analysis', 246.
33   Herring, 'Web Content Analysis', 245
34   Krippendorff, *Content Analysis,* 2004, 20.

In her critique of McMillan's five-step research protocol, Herring argues that web content analysis follows 'somewhat different norms from those traditionally prescribed for the analysis of communication content by researchers such as Krippendorff and McMillan' and may even be developing new norms.[35] She stresses that Krippendorff's framework also has been used rather liberally in content analysis practices. Furthermore, she notes, 'a growing number of web studies analyze types of content that differ from those usually studied in CA — such as textual conversations and hyperlinks — using methodological paradigms other than traditional CA'.[36] Herring offers a new list of five steps for web content analysis, or more specifically that of 'computer-mediated discourse analysis' (CMDA), which she initially developed in 2004.[37] CMDA is described as 'language-focused content analysis supplemented by a toolkit of spoken conversation and written text analysis'.[38]

Herring's checklist for web content analysis is similar to that of McMillan but offers in her view a more 'pragmatic' point of departure:[39]

a) Articulate research question(s);
b) Select computer-mediated data sample;
c) Operationalize key concept(s) in terms of discourse features;
d) Apply method(s) of analysis to data sample;
e) Interpret results.

Like Krippendorff and McMillan, Herring does not begin her procedures for content analysis with any specific mentioning of the exact means of collecting data but instead takes the data set to be something already given. Although the checklist may suggest that the research question would lead the analysis, at the same time she urges researchers to 'choose a research question that is "empirically answerable from the available data"'.[40] Herring also promotes flexibility in determining the sample types and coding categories based on the available data set. She builds a plea for a widening of the paradigms of content analysis, including the objects of such analyses, based on the assertion that most preceding approaches to content analysis focus on features and themes alone. She finds that in her own research practice of computer-mediated discourse analysis as applied to blogs, the research techniques of content analysis are indeed 'well suited for analyzing structural features of blog interfaces' and 'analyzing themes represented in blog entries and comments'.[41] Furthermore, although Herring rightly points out that the field of web content analysis nowadays extends beyond the use of conventional pre-web content analysis methods being merely applied to the web, it is clear that even this multi-disciplinary approach still attempts to separate content from its carrier. In this book, working beyond these concerted but insufficient attempts to update content

---

35   Herring, 'Web Content Analysis', 237.
36   Herring, 'Web Content Analysis', 238.
37   Herring, 'Web Content Analysis', 238.
38   Herring, 'Web Content Analysis', 238.
39   Herring, 'Web Content Analysis', 238.
40   Herring, 'Web Content Analysis', 238.
41   Herring, 'Web Content Analysis', 241.

analysis for the (changing) present media age, I want to show how and why this (persistent) separation of content and carrier can no longer hold with online networked content.

Technicity of Content

As outlined in the introduction to this chapter, my emphasis on the technicity of content stems from the observation that web content is networked. The networked character of online content means that content now includes technical agents that network it, such as in-text hyperlinks, tags, and social buttons.[42] Re-considering the early disciplined approaches of content analysis, we can see how networked content raises numerous methodological questions, many of which have been pointed out already, for instance in the above work of McMillan and Herring. When demarcating and collecting the relevant content at stake in analysis, one may wonder, for instance, where exactly the content of an article in an online newspaper ends. Should the hyperlinked pages be included in the study? How should social buttons be treated? Are all these links and buttons mere features to be counted and quantified, or should they be analyzed otherwise?[43] My propositions for networked content analysis urge the analyst to move beyond the analysis of web page features to treat the particular technicities of content — exactly this complexity — as part of the *text* under study, as Krippendorff would phrase this. Only when we include these technical specificities in the analysis of content rather than attempting to separate content from its carrier, can we meaningfully apply still-key foundational content analysis techniques to natively digital content.

In line with Krippendorff, who states that the meaning of content emerges *through its analysis*, we could say here that the technicity of the content, and further, the algorithmic logic behind platforms (such as Twitter and Facebook) and search engines (like Google) that rank and organize content, both serve and give shape to this technicity while forming the unique *context* of web content. The fact that online content is networked and dynamic shapes the context, and in turn, the means of the analysis. In the last part of this chapter, I will give an example of technicities of content from the platforms I study through Digital Methods, methods in which the quantitative measures that are built into the medium are deployed for networked content analysis. Krippendorff's sensitivity to the context of the text and the materiality thereof, which I observe to have receded in later content analysis methods formulations by scholars like McMillan and Herring, can from this point regain prominence for a networked content analysis.

---

42   This term 'agents' implies that these pieces of content have agency, which I argue is indeed the case. These technical specificities not only present or structure text differently, they are also co-authoring the text. The chapter on Wikipedia will provide examples of this, when I zoom in on the activity of software robots authoring and editing articles.

43   Similar questions arise in the research (and practice) of web archiving, where national libraries and other organizations aim to demarcate and archive a 'national web'. Similarly, internet censorship tries to demarcate 'forbidden content', and grapples with similar questions (see our study on the Iranian web: R. Rogers, E. Weltevrede, S. Niederer, and E. Borra, 'National Web Studies: The case of Iran', in J. Hartley, J. Burgess and A. Bruns (eds) *Blackwell Companion to New Media Dynamics*, Oxford: Blackwell, 2013, pp. 142-166.).

There is no single common *type* of online content, as we have seen from McMillan and Herring's attempt at an overview, alongside many other attempts, and as is evident from the examples of different types of web content I provide in the case studies that follow.[44] Rather than emphasizing the pluriformity of the web's content 'types', I would like to *productively* distinguish between various platforms with which content analysis must come to terms.[45] Platforms are 'portals or applications that offer specific Internet services, frameworks for social interaction, or interfaces to access other networked communications and information distribution systems'.[46] Many researchers have described how the Internet can easily be observed to be changing into a constellation of platforms, which are fast becoming our main means of accessing online information.[47] This tendency adds to the urgency for content analysis approaches to be able to deal with platform-specific aspects of content.

The approach of networked content analysis that I put forward, given these above considerations, is based on two overarching principles. The first is that web content is increasingly accessed and organized through search engines and platforms. The second principle is that the technicity of content should be part of the analysis of such networked content proposed. In this way, I consider techniques of content analysis that are inclusive of the specificity of the *platform* in networked content analysis, and that enable the researcher to study content, with an enhanced literacy for its dimensionality and movement, within and through the technical specificities and cultures of online content in context. This entails analytical sensitivity that recognizes that each platform networks, handles and serves content differently, for instance, search engines serving search results in a ranked list, Wikipedia cleaning and organizing its content with robots, and Twitter linking content through hashtags.

Networked Content Analysis With (or as) Digital Methods

Perhaps the most significant difference of emphasis, also from Krippendorf, that I am making in the proposition for networked content analysis is for the research question to lead to the collection of data or to a specific query within an existing data set, rather than the other way around. To emphasize this research need on the level of methodology and protocol is clearly quite contrary to the pre- and early-digital methods of content analysis (as shown in the research protocols earlier in this chapter). Networked content analysis can start with a question involving a set of actors in a specific issue, as I engage climate change skeptics (detailed in Chapter 3), and in a Latourian way follow these actors across platforms and sources, looking at their resonance, their language, and their networks. Such an approach to online content is partly drawing on the techniques and strengths of issue mapping, the multidisciplinary research practice described in the Introduction, where the objects of study

---

44   For an example of another attempt, see e.g., Weare and Lin, 'Content Analysis of the World Wide Web'.
45   The idea of content segmentation has been popular in Internet marketing since the early 2000s, where it refers to the segmentation of content within one website, to attract various audiences. See for instance 'Content Segmentation: Differentiate Your Brand Online', 5 April 2012, http://contentmarketinginstitute. com/2012/04/use-content-segmentation-to-differentiate-your-brand/.
46   Platform Politics, '*Platform Politics: Call for Papers: A Multidisciplinary Conference*', Cambridge, UK, 2011, http://www.networkpolitics.org/content/platform-politics-call-papers.
47   Helmond, *The Web as Platform*.

are 'issues' themselves, and where analysis may include how these issues manifest online, within specific platforms. Issue mapping can follow a topic as it traverses sources, for example, or capture multiple online spaces in a comparative analysis. An example of this is offered by Climaps, where a mapping of the issue of climate change across sources and platforms resulted in an online atlas of climate change adaptation.[48]

Given that this demarcation of content is such an essential part of networked content analysis research, much attention needs to be paid to the design and fine-tuning of search strings when using engines and related tools. Clarifying refined queries for specific source sets enables the researcher to answer the research questions with their gathered data. Rather than using predefined categories or translating jargon into more familiar terms, such inquiries also aspire to 'follow the actors' in their own (issue).[49] Thus, research queries respect the terms employed by the actors. Source sets may be conventional, such as from leading environmental or human rights organizations' public data, or they may be derived more directly from web engines or platforms, e.g., the leading organizations according to a search engine query, or the sub-issues resonating in a set of tweets.

Critical views on issue mapping with digital methods highlight the problem of the methods' and tools' dependency on already problematic proprietary walled gardens, and otherwise volatile ever-innovating commercial web platforms, such as Facebook and Twitter.[50] Scholars particularly warn of the sheer impossibility of distinguishing between the working logic of web platforms and the exemplarity of 'platform artifacts'.[51][52] For example, the most 'retweeted' content on Twitter might be the most Twitter-friendly content; therefore, we may only be finding out more about the logic of the platform itself, rather than the issue under study or the eventfulness of a particular tweet.[53] Consequently, when dealing with online content, we need to take into account the socio-technical logic of the platform itself as part of any analysis.[54] In fact, with the explosive rise of (big) data, attention to socio-technical logics of platforms must be further prioritized as social research increasingly makes use of what is called *Live Research*, where masses of content (with specific forms and technicities) are aggregated in real-time, copied onto other networks, and archived across the (social) web.[55][56] Furthermore, data analysis and the tools that enable this are built on highly dynamic web services. In a critique of the famous Google Flu Trends project, David Lazer et al. write how Twitter, Facebook, Google, and the Internet more generally are continually changing because of the actions of millions of engineers and consumers.[57] Understanding and studying these platforms as socio-technical

48   EMAPS, 'Climaps: A Global Issue Atlas of Climate Change Adaptation', 2014, http://climaps.eu/.
49   Latour, *Reassembling the Social*.
50   J. van Dijck, *The Culture of Connectivity: A Critical History of Social Media*, New York, NY: Oxford University Press, 2013.
51   Marres, 'Why Map Issues?'.
52   N. Marres and E. Weltevrede, 'Scraping the Social? Issues in Live Social research', *Journal of Cultural Economy* 6.3 (2013): 313–335, Rogers, *Digital Methods*.
53   Marres, 'Why Map Issues?'
54   Niederer and van Dijck, 'Wisdom of the Crowd or Technicity of Content?'
55   Back et al. 'Doing Real Time Research'.
56   Marres and Weltevrede, 'Scraping the Social?'
57   D. Lazer, R. Kennedy, G. King, and A. Vespignani, 'The Parable of Google Flu: Traps in Big Data

systems for what they are, is of utmost importance, as they are 'increasingly embedded in our societies'.[58] In this book, I develop such a socio-technological perspective on the controversy surrounding climate change as presented and debated on the web.

Consequent to the process of data collection, and then the querying of that data through refined search queries, the decision to visualize the data arises. Visualization here is not a mandatory step in the analysis. However, it can be considered an applied tool for the purpose of visual and descriptive analysis. While the 'descriptive turn' has been 'embraced' in contemporary sociology, it does come with its own ethical questions, if you will.[59] Each time a map is made, the researcher has to consider the appropriate output of the analysis 'in ranked lists, in cluster graphs, in line graphs, in clouds, on maps' and on a more abstract level, the visual, critical and even political aspects of map-making in their work.[60][61] Sociologist Tommaso Venturini, when discussing controversy maps, has described social maps as a visual interface to complex issues: 'To be of any use, social maps have to be less confused and convoluted than collective disputes. They cannot just mirror the complexity of controversies: they have to make such complexity legible.'[62] Similarly, visualization of data layered onto a geographic map of an area should render legible the complexity of the area, as well as the ways in which the social media platforms from which the data is taken from actually deal with that geo-location. It must be constantly borne in mind that map-based visualizations have been criticized for their oversimplification and reductionist approach to vast and multifarious data, highlighting some information and obfuscating other data for the sake of creating a 'display [of] what we already know'.[63]

I would, therefore, like to stress that in this book and in related research, the practice and objects of mapping are not efforts to ignore the distributed nature of today's technologies or data, or the richness of public debate, but in fact to gain a better understanding of the complex patterns and intersections of competing technologies as they intertwine form and content.[64] These maps then function as a navigational tool through a complex debate, rather than aiming at a reductionist narrative.[65] The map is neither the end product nor an aesthetic

Analysis', *Science* 343 (2014): 1205.

58   Lazer et al. 'The Parable of Google Flu', 1205.

59   M. Savage, 'Contemporary Sociology and the Challenge of Descriptive Assemblage', *European Journal of Social Theory* 12.1 (2009): 158. In this article, Savage makes a strong case for visualization research, stating that 'there needs to be more sociological interest in visualization as process, social artifact, and research tool'.

60   Digital Methods Initiative, *DMIR Unit #5: Cross-Platform Analysis*, Amsterdam: University of Amsterdam, 2015.

61   D. Wood and J. Fels, *The Power of Maps*, New York: The Guilford Press, 1992.

62   T. Venturini, 'Building on Faults: How to Represent Controversies with Digital Methods', *Public Understanding of Science* 21.7 (2010): 797.

63   G. Lovink, *Social Media Abyss: Critical Internet Cultures and the Force of Negation*, Cambridge, UK: Polity Press, 2016, 152.

64   See also S. Niederer, G. Colombo, M. Mauri, and M. Azzi, 'Street-Level City Analytics: Mapping the Amsterdam Knowledge Mile' in *Hybrid City 2015: Data to the People*, Athens: University of Athens, 2015, www.media.uoa.gr/hybridcity.

65   As some of my research was part of the EMAPS EU-project (2014), I'd like to refer here the way in which the analysts of the program articulated their practice of mapping while showing full awareness

inquiry into the data. Here, the visualizations function as an analytical tool.[66] The maps then enable researchers to essentially zoom out, navigate the issue, and decide the directions for further analysis. I endeavor to accomplish this here by studying a single issue across multiple platforms from different viewpoints and by creating not one all-encompassing 'mother map', but a series of different maps and descriptions, of variegated utility, which underline both the complexity of studying issues through online content and the entanglement of content with its technicity. Of course, offline mass media content will also be present in these maps and analyses, as news and other media (which have traditionally been subject to content analysis) are referred to in and thus form part of online networked content.

When applied to the study of controversy, as in this case the climate change debate, the key contribution of networked content analysis lies in the development of adaptive research techniques that are rooted in content analysis while suited to networked digital media content. These methods allow researchers to follow debates and actors across diverse sources and online platforms. In the next chapters, I will operationalize such an approach, in which I discuss first how content is networked (on the web and accessed through Google, Wikipedia and, lastly, Twitter) and then address a question considering a specific aspect of the climate change debate. In the case of the web, I assess the place and status of climate change skeptics, within climate science and on the web. Are they professional climate experts, or professional skeptics? I operationalize this question by taking a set of key actors and profiling them, if you will, by assessing their prominence within climate science, their networking behavior, their resonance in search engine results for the issue of climate change and, lastly, by appraising and discussing their 'related content'. In the case of Wikipedia, a network of interlinked articles on climate change and global warming allows for a reconstruction of the debate over time. Lastly, through the platform of Twitter, I provide a comparative view of the different stages of climate change (skepticism, mitigation, adaptation, and vulnerability/conflict), and explore the sub-issue of climate vulnerability in detail.

Conclusions

Content analysis has made longstanding contributions to the broadest definitions of mediated 'textual' analysis, but when applied to networked content evokes a myriad of analytical issues: demarcating the object of study (where does a website end?), dealing with the dynamic character of the web (how can you redo the research, when the object of study constantly changes?), dealing with the unknown algorithms of search engines (how does one rely on Google without knowing its exact algorithm?) and so on. Where some content analysts, such as McMillan, prefer to stay close to the foundations of content analysis, others, such as Herring, make a plea for the widening of this research paradigm and its object of study, through the

---

of the value of these emerging critiques, which can be read in *Climaps*, the collaborative issue atlas of climate change adaptation produced as the result of EU FP7 project EMAPS (with principal investigator Bruno Latour, Sciences Po) and Digital Methods at the University of Amsterdam collaborated with international parties (Barcelona Media, Politecnico di Milano, the Young Foundation, and the Dortmund Technische Universität) in mapping the issue of climate adaptation. EMAPS, 'Climaps.'

66   S.K. Card, J.D. Mackinlay, and B. Shneiderman, *Readings in Information Visualization: Using Vision to Think*, San Francisco, CA: Morgan Kaufmann, 1999.

inclusion of methods from adjoining scientific fields. However, while Herring regards content as contained in media documents, I argue that the separation between content and its carrier no longer holds with networked content.

As Krippendorff pointed out, it is the specificity of the definition of 'content' one chooses that leads to specific kinds or varieties of content analysis.[67] The inclusion of web content's technicity into the idea of content itself leads to analyses that make use of and deal analytically with these technical agents. As I have demonstrated in this chapter, the collection and analysis of web content that follow the specificities of each platform and operationalizes the specific technicities at play will lead to more precise analysis, one that is sensitive to the networked nature and dynamical movement of online content. I have here realigned my work with Krippendorff's inceptive call to keep the content together with its carrier (or context), and accordingly propose that in networked content analysis researchers include not only the carrier (e.g., the Wikipedia article, the search engine result, the Tweet) but also the technicity (e.g., the editing history and content robots of the Wikipedia article, the ranking of the search results, the hashtags and retweets networking the collection of tweets) as part of their analytical approaches.

For the collection of content in early instituted content analysis methods, the data available was always shaping or assumed to be setting up the point of departure for research. In other words, research questions enabled the researchers to sample more specific queries from that available data only. In the methods proposed here, I assert the value of working the other way around. The collection of data occurs *after* the research questions are formulated, and starts with the careful composition of source lists that are to be queried. After the sources are collected, and the spreadsheets are in place, the queries for the content sphere's dominant engine are designed, tested and, if necessary, tweaked. Subsequent analysis of the content under study often comes with a map or visualization of the data.

The way forward presented here is a first step in the description of the contribution of medium-specific digital methods to the field of content analysis. Of course, it will need and welcome further elaboration, and, to stay in line with traditions of content analysis, should offer both a description of the approach on a theoretical, conceptual and historical level and eventually also hands-on guidelines that lay out the recipe for a solid project of content analysis. Clearly, I am valuing and making progress towards tools and methods for networked content analysis that stay tied to the inceptive work of Krippendorff. In line with his thinking, a contemporary web-literate approach titled networked content analysis remains open to all kinds of content and includes contents' technical specificities in the value of such. The case studies in the next chapters offer such medium-specific approaches to climate change content on the web (and Google), Wikipedia, and Twitter to ask which methods might be further tailored towards platform-specific ends, and which can be scaled from or between platforms.

---

67   Krippendorff, *Content Analysis*, 2004.

3. CLIMATE DEBATE ACTORS IN SCIENCE AND ON THE WEB

On 12 December 2015, a consequential agreement in the history of global climate negotiations was reached when 195 countries adopted the so-called *Paris Climate Agreement* during the 21st annual Conference of the Parties, better known as COP21. Two weeks of 'fierce negotiations' ended with the words 'I hear no objection in the room, I declare the Paris Climate Agreement adopted', spoken by the president Laurent Fabius.[1] Loud cheers followed, and festive pictures were published along with a dedicated hashtag #ParisAgreement (Figure 2).

Figure 2: #ParisAgreement. Cheers after the declaration of the adoption of the Paris Climate Agreement on December 12th of 2015.[2]

The global agreement that was adopted is a substantiation of the widespread consensus on climate change as a most urgent issue of our times, or as it is phrased in the agreement itself, of:

> [R]ecognizing that climate change represents an urgent and potentially irreversible threat to human societies and the planet and thus requires the widest possible cooperation by all countries, and their participation in an effective and appropriated-international response, with a view to accelerating the reduction of global greenhouse gas emissions.[3]

1   United Nations Conference on Climate Change, 'COP21', 2015, http://www.cop21.gouv.fr/en/.
2   United Nations Framework Convention on Climate Change, 'Adoption of the Paris Agreement', United Nations, 12 December 2015, https://unfccc.int/resource/docs/2015/cop21/eng/l09.pdf.
3   United Nations Framework Convention on Climate Change, 'Adoption of the Paris Agreement', 1.

As such, the agreement marks an important milestone in the climate debate, which has spread from a scientific debate to a public debate. Arguably, the agreement may also signal a new chapter, indicating that, in fact, skepticism is losing ground. In the case studies that follow, I begin tracing 'new' or unfolding elements of this debate from 2008, the date of the first international skeptics' conference, as discussed in the Introduction, moving all the way to 2015. However, engaging the technicity of web content (i.e., that of Wikipedia) sometimes requires the researcher to go back in time to allow for a historical reconstruction of present issues or to assess earlier milestones in a current controversy, such as important IPCC reports and COP events before 2008. As this book is not a historiography of the debate but rather a study of the controversy through networked content, I will furthermore not always discuss the events in chronological order. In my research, tracing the climate change controversy involves encountering certain objects, images, publications, or events that resonate strongly or even cause a heating up of the debate.

The most famous climate controversy object, in my view, is the so-called 'hockey stick graph,' a chart in which a thick black line powerfully depicts a sharp and unprecedented rise in global temperatures since the late 20th century (Figure 3, p. 42). The hockey stick graph has been widely published, for instance, in the IPCC report of 2001. But a wider audience may know it from the performative account of climate change given in the documentary *An Inconvenient Truth* (2006), in which Al Gore projects the graph and uses a lift to follow the unprecedented rises in temperature and CO_2 levels all the way toward the top of the screen. In 2009, the hockey stick graph found itself at the center of the climate debate again, with the so-called Climategate scandal. Following a hack of East Anglia University's climatic research unit, a selection of emails was leaked in which climate scientists described the making of the hockey stick graph for publication in the journal *Nature* as the fraudulent-sounding 'Mike's Nature Trick', in which research visualizations 'leave out the anomalies'.[4][5]

Here rather than contributing to a critical discussion of climate science, which is by no means my area of expertise, I ask how content analysis may be amended to include networked content's technicities, and, in doing so, learn from controversy analysis and digital methods. This book thus aims to contribute to those respective fields, as well as to previous scholarly work on the climate debate, especially its mediation through networks. Academic research on the climate debate has taken as its point of departure the histories of events, objects, and scandals, and studies their coverage, framing and impact across a broad spectrum of mediation, from mass media to scientific literature. Scholars have, for instance, focused on public awareness and the general public's engagement with the issue.

---

4   F. Pearce, 'The Five Key Leaked Emails From UEA's Climatic Research Unit', *The Guardian*, 7 July 2010, http://www.theguardian.com/environment/2010/jul/07/hacked-climate-emails-analysis.

5   However, research based on Google Trends data has shown that the Climategate scandal, in retrospect, has had an only short-lived effect on the public debate around climate change. W.R. Anderegg and G.R. Goldsmith, 'Public Interest in Climate Change Over the Past Decade and the Effects of the "Climategate" Media Event', *Environmental Research Letters*, 9.5 (2014): 054005.

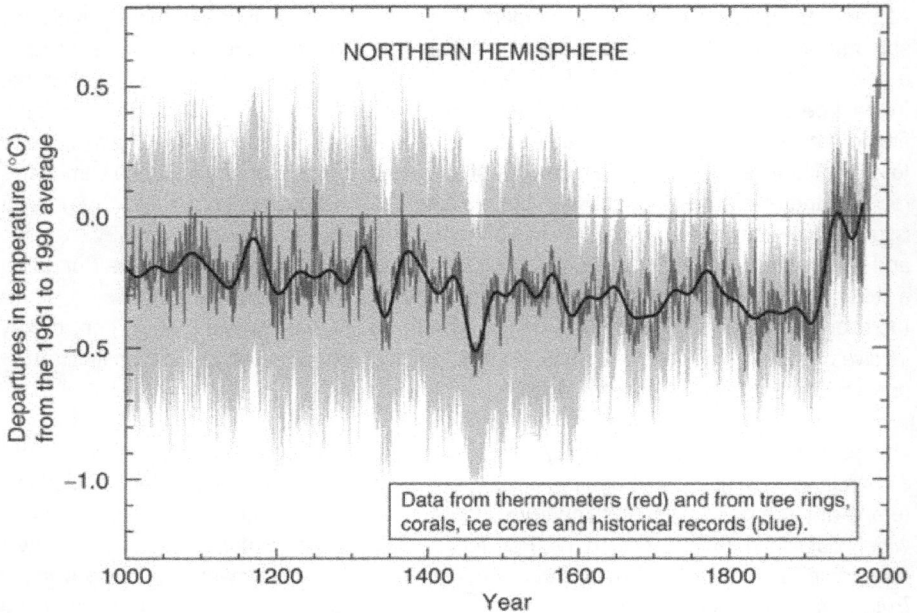

Figure 3: The HockeyStick Graph. Graph showing unprecedented rises in temperature since the late 20th century.[6]

With central questions such as 'Do people believe in climate change? And is the percentage of people who believe climate change is taking place increasing or decreasing?' surveys and polls are undertaken organizations such as Pew Research Center.[7] The outcomes of these reports are referenced in scholarly works that look at public opinion and the public understanding of climate change.[8] Other scholars have created timelines and so-called trend chronologies, which 'summarize public opinion across key dimensions including [...] public awareness of the issue of global warming' to analyze the development of public opinion over time.[9] Another strong tradition of climate debate research in the tradition of media monitoring to measure the coverage of the issue in the news, for instance, by comparing its coverage across a set of newspapers.[10]

6   M.E. Mann, R.S. Bradley, and M.K. Hughes, 'Northern Hemisphere Temperatures During the Past
     Millennium: Inferences, Uncertainties, and Limitations', *Geophysical Research Letters* 26.6 (1999):
     759–762.
7   A. Kohut, D.C. Doherty, M. Dimock, M., and S. Keeter, 'Fewer Americans See Solid Evidence of Global
     Warming', *Washington, DC: Pew Research Center,* 2009.
8   S.C. Moser, 'Costly Knowledge – Unaffordable Denial: The Politics of Public Understanding and
     Engagement on Climate Change', in *The Politics of Climate Change: A Survey,* 2010, 155–181.
9   Nisbet, M.C. and T. Myers, T. 'The Polls-Trends: Twenty Years of Public Opinion About Global Warming',
     Public Opinion Quarterly 71.3 (2007): 444.
10   Djerf-Pierre, 'When Attention Drives Attention'.

A different strand of climate debate coverage research longitudinally monitors the coverage of the climate debate in mass media.[11] Here, the focus can be on television shows or printed news, or on specific features of the coverage, such as the use of imagery in environmental news.[12] Longitudinal analysis of news coverage can reveal the so-called 'issue attention cycles' in a specific country or in a comparison across countries.[13] The related concept of *news spirals* refers to the phenomenon that once the climate is in the news, this creates a general upsurge of other environmental news.[14] Studies specifically centered on events and scandals zoom in on controversy objects such as Climategate or debates around (alleged mistakes in) the IPCC reports.[15][16][17] Rather than looking at controversy objects as a starting point, in the following case study, I will enter the climate debate through the scope of its actors, who I will approach using scientometric analysis and networked content analysis, for which I will conduct both hyperlink analysis and search engine-based resonance analysis.

As the climate debate is not limited to a single communication channel but takes place across online platforms, I will first consider Google Web Search — a dominant entry-point to the web for many — as a beginning platform through which to operationalize some endeavors of capturing, reading and analyzing this controversy's content. Whereas Google Web Search has grown dramatically since 2008 (as have Twitter and Wikipedia, the other platforms discussed in the next case studies of Chapters 4 and 5), its role in controversies has not been systematically examined. In this first case study, I will discuss Google Web Search in a Networked Content Analysis of the climate controversy in the period of 2008 - 2011. The case study asks how the technicities of networking (through hyperlinked websites) and search (e.g., its output of ranked lists) might be used to measure the prominence of specific actors in specific issues, in this case, looking at the networks and resonance of climate change actors. By 'climate change actors', I mean to indicate both non-skeptical climate scientists (for lack of a better term) and climate change skeptics. 'Climate change skeptics' here refers to those skeptical of climate change and its sub-issues such as human-made global warming, unprecedented global warming (temperature rises), and a variety of the methods employed to study climate change. All scientists are 'skeptical' to a certain extent, so when I use the term 'non-skeptical climate scientists', it refers to scientists who do not publish skeptical articles on the anthropogenic causes or unprecedented effects of climate change. I choose the term skeptic over

11   A. Nacu-Schmidt, K. Andrews, M. Boykoff, M. Daly, L. Gifford, G. Luedecke, and L. McAllister, 'World Newspaper Coverage of Climate Change or Global Warming, 2004-2016', 2016, http://sciencepolicy. colorado.edu/media_coverage.
12   S.J. O'Neill, M. Boykoff, S. Niemeyer, and S.A. Day, 'On the Use of Imagery for Climate Change Engagement', *Global Environmental Change* 23.2 (2013): 413–421.
13   D. Brossard, J. Shanahan, and K. McComas. 'Are Issue-cycles Culturally Constructed? A Comparison of French and American Coverage of Global Climate Change', *Mass Communication & Society*, 7.3 (2004): 359–377.
14   Djerf-Pierre, 'When Attention Drives Attention'.
15   Anderegg and Goldsmith, 'Public Interest in Climate Change Over the Past Decade and the Effects of the "Climategate" Media Event.'
16   A.J. Hoffman, 'Talking Past Each Other? Cultural Framing of Skeptical and Convinced Logics in the Climate Change debate', *Organization Environment* 24.1 (2011): 3–33.
17   B. Nerlich, '"Climategate": Paradoxical Metaphors and Political Paralysis', *Environmental Values* 19.4 (2010): 419–442.

'denialist' (a stronger term often used by those who stand in opposition to these skeptical actors) while bearing in mind that the term 'alarmist' as used by climate change skeptics to describe their opposition is also rhetorically overloaded. Importantly, 'deniers' and 'alarmists' are labels used by others to define and already delegitimize these specific actors, not by the actors to describe themselves.

To assess what the techniques of networked content analysis may add to the study of the climate controversy, I pair its approach with that of scientometrics (or the quantitative study of science), a traditional means to study the prominence of scientific actors within a specific scientific field.[18] In all, this chapter assembles a profile of these actors' (aspired to) positions inside and outside academia and offers a finer-grained picture of the status, group formations and issue commitments of climate change skeptics. The chapter in this way assumes that the *question* of whether these actors are scientists or lobbyists holding open or reopening the climate debate into controversy is an extremely current question, and that finding answers towards such questions as I do here, is integral to a better understanding of the climate change debate's entanglement with stakeholders.

In the Introduction, I have outlined that the group formation of these skeptics has been key to climate change becoming a major controversy. As I detailed in that chapter, the first international conference for climate change skeptics was organized in March 2008. The Heartland Institute, a Chicago-based libertarian public policy think-tank, organized this event with the inaugural title, *Can You Hear Us Now? Global Warming Is Not a Crisis!* The format was that of a traditional scientific conference with three days of parallel sessions and keynote speakers as well as online proceedings.[19] In his opening remarks, Heartland's president Joseph L. Bast stressed that the conference featured talks by 'over 200 scientists and other experts from leading universities and organizations from all over the world'. Bast furthermore stated that:

> These scientists and economists have been published thousands of times in the world's leading scientific journals and have written hundreds of books. If you call this the fringe, where's the center?[20]

Bast gave credence here to climate skeptics as core actors in climate science, while most descriptions of climate change skeptics, whether by watchdogs (e.g., watchdogs of corporate PR campaigns such as SourceWatch.org), journalists or scientific analysts, paint a less flattering picture. Scholars have emphasized how skeptics effectively keep the climate conversation alive *as a controversy* in the face of increasing statements of consensus from

---

18   This study was published in the European Journal of Media Studies, *NECSUS*. S. Niederer, 'Global Warming Is Not a Crisis! Studying Climate Change Skepticism on the Web', *Necsus* 3 (Spring 2013): http://www.necsus-ejms.org/global-warming-is-not-a-crisis-studying-climate-change-skepticism-on-the-web/.
19   The Heartland Institute, 'First International Conference on Climate Change'.
20   The Heartland Institute, 'First International Conference on Climate Change'.

the global scientific climatology community.[21] [22] Skeptics are often criticized for having strong ties to specific industries invested in the status quo reproduction of our climate-changing economy, as described in books such as *Merchants of Doubt* and *Doubt is their Product*, as well as the report *Smoke, Mirrors and Hot Air: How ExxonMobil Uses Big Tobacco's Tactics to Manufacture Uncertainty on Climate Change* and various academic papers.[23] [24] [25] These publications describe how industry-funded skeptics insist on the lack of consensus on anthropogenic (i.e., human-induced) global warming, using strategies from prior decades' tobacco industry-funded research that downplayed truth claims on the health risks of smoking. In October of 2015, this topic flared up in the news, as the New York attorney general announced an investigation of Exxon Mobile 'to determine whether the company lied to the public about the risks of climate change'.[26]

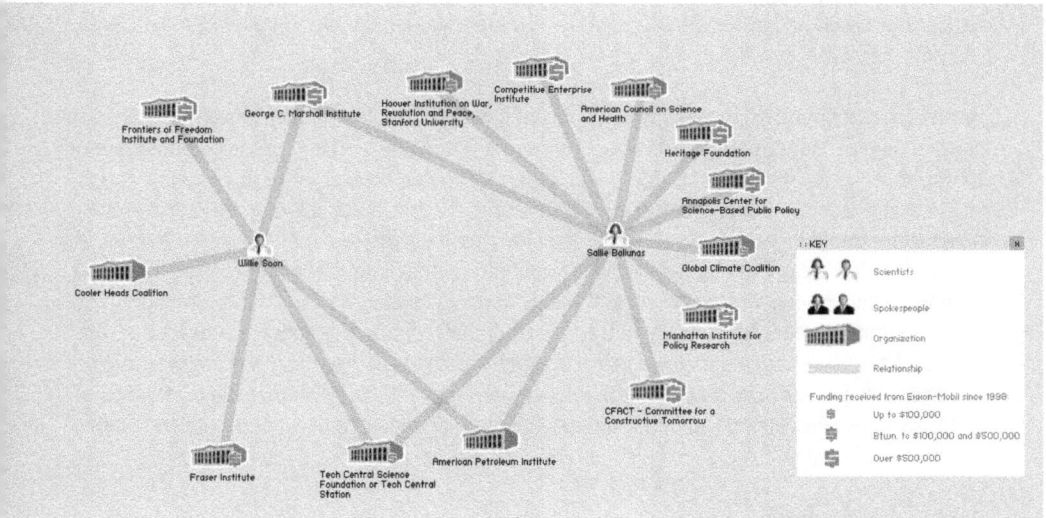

Figure 4: Exxonsecrets. Map showing the institutional relationships of Willie Soon (left) and Sallie Baliunas (right) and their funding by Exxon-Mobil since 1998.[27]

21 N. Oreskes, 'Beyond the Ivory Tower: The Scientific Consensus on Climate Change', *Science* 206.5702 (2007): 1686.
22 N. Oreskes, 'The Scientific Consensus on Climate Change: How Do We Know We're Not Wrong?' in J.F.C. DiMento and P. Doughman (eds) *Climate Change: What It Means for Us, Our Children, and Our Grandchildren*, Cambridge: MIT Press, 2007, pp. 65-99.
23 Michaels, *Doubts Is Their Product.*
24 Union of Concerned Scientists, 'Smoke, Mirrors and Hot Air.'
25 C.W. Schmidt, 'A Closer Look at Climate Change Skepticism', *Environmental Health Perspectives* 118.12 (2010): A536–A540.
26 J. Gillis and C. Krauss, 'Exxon Mobil Investigated for Possible Climate Change Lies by New York Attorney General', *The New York Times*, 5 November 2015, http://www.nytimes.com/2015/11/06/science/exxon-mobil-under-investigation-in-new-york-over-climate-statements.html.
27 'Greenpeace', http://www.exxonsecrets.org/maps.php?mapid=1804.

The industrial and financial ties of climate change actors have been visualized for public awareness and comprehension in projects such as Exxonsecrets. This watchdog project by Greenpeace shows key scientists, spokespersons, and organizations that have received Exxon-Mobil funding since 1998. Figure 4 (p. 45) shows a map of the affiliations of the prominent climate change skeptics Willie Soon and Sally Baliunas and depicts which of those organizations have received funding from ExxonMobil. On the left, Soon is depicted as having six institutional affiliations (for instance, with the George C. Marshall Institute and the Fraser Institute), four of which have received funding from ExxonMobil and one of which is the American Petroleum Institute. On the right-hand side, Baliunas is shown to hold eleven institutional relations, ten of which have received ExxonMobil money and one of which is also the American Petroleum Institute. Economic visualizations like this, of supposedly 'disinterested' scientific debate and controversies, are designed to activate public comprehension of bias and (sometimes artificial) controversy in the networked, public mediation of so-called 'scientific research' on climate change.

Rather than zooming in on the industrial ties of specific climate change researchers, in this chapter, I want to zoom out to laterally consider the place and status of climate change *skepticism*, that is, the resonance of skeptics, within the networked content of climate change science and its online debate. I will start with a brief discussion of scientometric analyses of these prominent climate skeptics.[28] In these analyses, I take to the opening statement of the first international climate change skeptics' conference abovementioned, to turn its claims of authority into a question. Putting aside the epistemological claims of the conference content, I trace its main actors to assess whether these climate change skeptics are indeed at the 'center' of climate change debates. Here, I drew on a data set of over 15,000 scientific articles on climate change that had been cited at least three times, to find whether these skeptics — speakers of the first Heartland conference — were indeed located at an authoritative 'center' of climate science.[29]

Related to this understanding of citation networks, the utility of hyperlinks for online content analysis has been asserted by many scholars. Here, I would like to point specifically to the work by those media scholars who describe links as being both an indicator of reputation and the performance of politics of association.[30][31] For example, not all organizations link to all other organizations that work in the same field; they rather link only to the organizations that

---

28   Some of the specific research methods that I am employing may be unfamiliar to existing content analysis or other media studies approaches. Scientometrics uses data sets of scientific publications and assesses these through citation analysis. More specifically, scientometric analyses can extend from tracking citational behavior and referencing, to understanding these processes as constructing norms and rules of scientific writing, to considering how specific or groups of texts play out in an interreferential network of influence and authority. Citational behavior as indexed by ISI Web of Science, thus provides the researcher with a searchable data set of scientific publications that are networked by interlinking. Wouters, *The Citation Culture*.

29   For this study that tested claims made in 2008, the sample is limited to those publications cited at least three times by July of 2008.

30   A. Dekker, *The Politics of Association on Display: Interview with Richard Rogers*, Amsterdam: Netherlands Media Art Institute, January 2008, http://nimk.nl/eng/the-politics-of-association-on-display.

31   Gerlitz and Helmond, 'The Like Economy'.

NETWORKED CONTENT ANALYSIS: THE CASE OF CLIMATE CHANGE   47

they prefer to be associated with. New media scholar Axel Bruns describes the IssueCrawler, the hyperlink analysis tool used in this chapter to conduct hyperlink analysis and visualize the hyperlink networks, as 'predominantly designed for identifying "issue networks", that is, networks of websites which form around the interlinking and exchange of information pertaining to specific issues or topics'.[32] This technique of hyperlink analysis has been applied to the climate debate before. In the paper *Landscaping Climate Change* (2000), Rogers and Marres describe the study of hyperlinking as a means to map the debate around an issue. They regard linking as a way to recognize other participants in the debate and '[s]imilarly, non-linking is a sign of non-recognition, or, more radically, is an act of silencing through inaction. (Greenpeace does not link to Shell, but Shell links to Greenpeace)'.[33] When thinking of a hyperlink in terms of recognition or as politics of association, the link can also be deemed and repurposed as an instance of group formation, as described in the Introduction in reference to the work of philosopher and anthropologist Bruno Latour, who has argued cogently for the fact that there are no groups 'without a rather large retinue of group makers, group talkers, and group holders'.[34] [35]

Important to mention here is that this use of hyperlink analysis has been recognized as an important technique by content analysis too. In the third edition of *Content Analysis: An Introduction to its Methodology* (2013), Klaus Krippendorff describes hyperlink analysis as a means to study issue networks and answer issue-related research questions regarding the composition of actors, influence and authority within the network, and the life of an issue over time (when conducting longitudinal analysis).[36] [37] [38] The mentioning of hyperlink analysis and its qualities for *issue research* in this handbook not only demonstrates again the willingness of content analysis in its original form to open up to such digital research methods and objects. The inclusion of this research technique also stresses the importance of the further development of networked content analysis and makes a case for the inclusion of hyperlink analysis therein. As hyperlinks are the basic 'webby' way to network online content, they are an essential means to trace and capture affiliations, aspirations, and alignment between actors.

The third way in which I will measure the reputation of climate actors and their viewpoints within larger contestations of climate change knowledge online is through what I call 'resonance analysis'. Here, a demarcated set of sources, in this case, the top results for the query of [climate change], is assessed for the presence (and absence) of climate change skeptics, as well as other scientists in the top search findings. This is of interest on two levels.

32   A. Bruns, 'Methodologies for Mapping the Political Blogosphere: Explorations Using the IssueCrawler Research Tool', *First Monday*, 12.5 (2007): http://firstmonday.org/ojs/index.php/fm/article/view/1834/.
33   Rogers and Marres, 'Landscaping Climate Change'.
34   Rogers, *Information Politics on the Web*, vii.
35   Latour, *Reassembling the Social*, 32.
36   Krippendorff, *Content Analysis*, 2013, 234-235.
37   Here Krippendorff refers to the definition by Heclo, who 'introduced the term in 1978 to describe connections between people who regard each other as knowledgeable and interested in particular public policy issues and who work these issues out essentially among themselves' Krippendorff, *Content Analysis*, 2013, 233.
38   Krippendorff also cites Rogers, who uses the term in reference to the output of the IssueCrawler. Krippendorff, *Content Analysis*, 2013, 234.

Firstly, we may ask which sources make it into the top of the results in Google Web Search. Often critically referred to as a 'black box', due to its undisclosed algorithm, it is known that Google grants status to sources that are both established (as in receiving many in-links from other websites) and relevant as in often clicked, a logic that has been discussed in relating PageRank to citation analysis.[39][40][41] Secondly, such analysis makes visible which sources grant a voice to the skeptics prior-identified from the Heartland Conference program. Which of the top 100 results are proven to be the most 'skeptic-friendly' websites? And who of the prior identified skeptics appear most frequently? This harkens back to traditional means of content analysis, where in media monitoring the airplay of specific actors (for instance, a 'Democratic presidential candidate' versus a 'Democratic presidential candidate in a televised pre-election debate') would be counted and analyzed. Lastly, it enables an assessment of who makes it into the top results (a technique referred to as 'source distance analysis').[42]

This actor-centric approach to the comprehension of the work and networks of skeptics leads me to raise further important questions about climate change skeptics' *issue commitment,* once it becomes possible – and useful – to map skeptics' (non-) scientific publications on topics other than climate change. In other words, their claims towards scientific rationality can be further researched by asking whether such claims are actually (*scientifically)* concerned with climate change at all, or with the political questions that acceptance of the science might raise. This is less of a radical move than it may seem at first glance. Consider, for example that one of the most prominent Dutch climate skeptics is the president of Stichting Skepsis, a foundation that deals with not just one controversy but many (making climate change a target of purposeful scrutiny to the point of delegitimization, alongside topics like homeopathy and so on). To understand how such controversies are networked online into *issue relations,* through issue actors is to understand the network of a controversy's content and actors, and to study the complex ecologies of debates through the distances and connections between

39   Lawrence Page, Sergey Brin, Rajeev Motwani, and Terry Winograd. 'The PageRank Citation Ranking: Bringing Order to the Web,' Technical Report, Stanford InfoLab, 1999, http://ilpubs.stanford. edu:8090/422/.

40   Rieder, 'What Is in PageRank?' In this paper, Rieder conducts a historical analysis of PageRank through two paper publications (S. Brin and L. Page, 'The Anatomy of a Large-Scale Hypertextual Web Search Engine', *Computer Networks*, 56.18 (2010): 3825–3833.; Page et al. 1999) and two US patents for PageRank, and explores their references to citation analysis (and similarly to sociometric literature), where the patents interestingly prove a richer resource for such references.

41   See also: E. Weltevrede, *Repurposing Digital Methods: The Research Affordances of Platforms and Engines*, University of Amsterdam, Amsterdam, 2016, 105, for her historical discussion of the changes in its algorithm over time which she bases on 'Page and Brin's whitepaper (1998), key patents and empirical projects' and in which she underlines that 'Google Web Search's current algorithm is not only PageRank but consists of over 200 signals and metrics'. Relevant to note here is that, as in my own work, Weltevrede strives not to *know the algorithm* but to *research with algorithms*. See also Clay Shirky's speculation on 'algorithmic authority', or the discussion of the trust people place in the algorithms of Google, Twitter and Wikipedia alike in Rogers's *Digital Methods*. C. Shirky, 'A Speculative Post on the Idea of Algorithmic Authority', 2009, http://www.shirky.com/weblog/2009/11/a-speculative-post-on-the-idea-of-algorithmic-authority/. Rogers, *Digital Methods,* 96.

42   Rogers, *Digital Methods,* 112.

them (as I will bring into practice again in chapters 4 and 5 on Wikipedia and Twitter). Such analyses give space to the *drama* in the network, to paraphrase Noortje Marres.[43]

Before applying these three methods of analysis to the actions and impacts of skeptics (through hyperlink analysis, actor-oriented actor resonance analysis, and actor-issue commitment analysis), a brief discussion of my mapping of the issue *within* science is necessary. Here I make use of the ISI Web of Science to chart the position of these skeptical scientists within climate science; from this, I can test how *fringe* or *central* to climate science these actors are.

Climate Change Skeptics: Mainstream or Fringe?

This scientometric analysis of the identified skeptics' position tests the claim that they are in the scientific center (of climate science). A question that might be raised is whether climate change skepticism should be considered to be its own field in the sense of a particular distribution of disciplines, and following this, whether the composition of climate skepticism mirrors that of climate science. In other words, whether they are doing the same science and generating different results that would technically add up to a 'controversy' or whether something more complex than this is operative in the politics of online climate change knowledge. Thus, to consider whether skeptics are at the 'center' of climate science, I will first compare the academic disciplines of skeptical authors and assess whether they mirror the composition of climate science authors. I then look at publications in climate science and compare these to a subset of academic publications from climate skeptics. The aim of this analysis is to first to get a better understanding of the place and status of skeptics within climate science, and then to complement this with networked content analysis techniques capable of shedding light on the role of such actors within a controversy that plays out inside and outside of advanced, scientifically adjudicated academic research settings.

The basic data for my starting point is a list I have compiled of prominent skeptics to which I will apply the scientometric analysis. The prominence of the actors has been determined through reference to prior-developed listings of climate change skeptics mentioned by Wikipedia entries, previously mentioned watchdogs and other scholars' academic analyses of climate skepticism.[44] Cross-referencing these existing listings with the line-up of keynote speakers at the Heartland conference of 2008, resulted in a shortlist of fifteen prominent climate change skeptics: Sallie Baliunas, Joseph Bast, Paul Driessen, William Gray, Sherwood Idso, Václáv Klaus, Richard Lindzen, Patrick Michaels, Steven Milloy, Frederick Seitz, S. Fred Singer, Willie Soon, Roy Spencer, John Stossel, and James M. Taylor. Concurrent to this assembling of prominent skeptics, I queried the ISI Web of Science for all articles on 'climate change'.

---

43 Marres, 'There is Drama in Networks'.
44 For the compilation of the list, I have triangulated lists of skeptics from: A.M. McCright and R.E. Dunlap, 'Defeating Kyoto: The Conservative Movement's Impact on US Climate Change Policy', *Social Problems* 50.3 (2003): 348–373; Mother Jones, 'Put a Tiger in Your Think Tank,' *Mother Jones,* 2005, http://www.motherjones.com/politics/2005/05/put-tiger-your-think-tank; Sourcewatch, n.d.; Wikimedia contributors, 'Bot Activity Matrix', http://stats.wikimedia.org/EN/BotActivityMatrix.htm. Frederick Seitz passed away prior to the conference yet has been kept on the list.

On 9 July 2008, there were approximately 27,000 articles, 15,877 of which received at least three citations; these form the list of articles retained for the analysis.

Using this data set of nearly 16,000 articles with the list of skeptics, I compare the disciplines of the journals in which significant climate change articles appear to the disciplinary backgrounds of the climate skeptics and their co-authors. This first analysis demonstrates that seven out of the top 10 disciplines in the climate sciences are present in the skeptics' top 10: ecology, meteorology, and atmospheric sciences, multidisciplinary sciences, environmental sciences, interdisciplinary geosciences, plant sciences, and agronomy. The climate change skeptics' disciplinary composition partially matches that of climate science, besides having some signature disciplines of its own (within the top 10 most occurring disciplines), namely astronomy and astrophysics, biochemistry and molecular biology, and medicinal chemistry. Disciplines unique to the rest of climate science are multidisciplinary sciences, forestry, and environmental engineering. These large overlap in the disciplinary background of the climate scientists publishing (cited papers) on climate science and the subset of climate skeptics seems to confirm Bast's statement that skeptical climate science is, in fact, part of climate science and not positioned outside the field. Or, at least, it resembles climate science in terms of the composition of scientific disciplines involved. Knowing the place of climate change skeptics within the climate science disciplines, I now want to test whether the skeptics publish in prominent climate science journals or whether they have their own dedicated skeptics' journals.

Using the ISI result files and ReseauLu (the network analysis software), I compare which journals *do not* publish skeptics at all, which publish *only* skeptics, and which journals publish *both* skeptics as well as non-skeptical views. Here it is found that the shortlisted climate skeptics and their co-authors publish in the top four climate journals (which are in the shared nodes in the center of the network). This may be counterintuitive, especially when thinking about the aforementioned readings of the climate change skeptics' 'lobby' in which these actors are described as a relatively small but powerful group of scientists of which 'the most vocal skeptics were *not* qualified, were *not* working in the field'.[45]

Figure 5 (p. 51) shows the visualization of the results. In the center are the shared nodes. These are the 30 publications that publish articles (cited at least three times) by skeptics as well as others. The shared journals include prominent academic publications, including *Nature, Science, Journal of Climate, Geophysical Research Letters, Journal of Geophysical Research,* and *Climatic Change*. This is where climate change skepticism overlaps or resides within the rest of climate science. On the left are the journals that do not publish work by our short-listed skeptics and their co-authors. On the right are the nodes that represent the journals that publish only the works of climate change skeptics.

---

45   J. Hoggan and R. Littlemore, *Climate Cover-up: The Crusade to Deny Global Warming,* Vancouver: Greystone Books, 2009, 4.

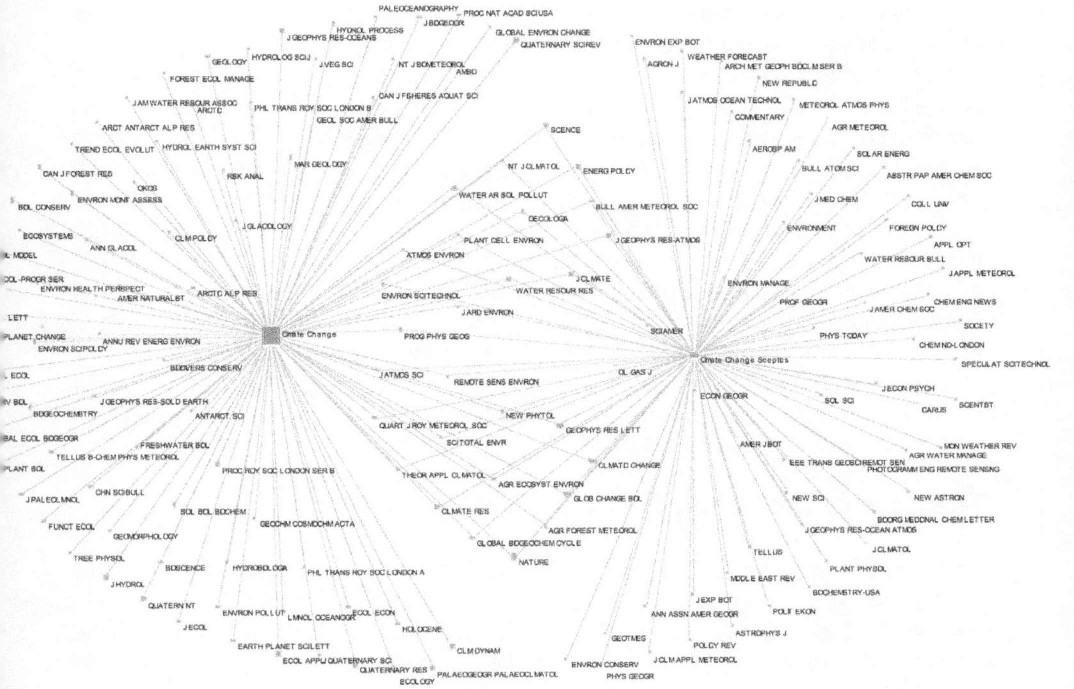

Figure 5: Climate science publication graph. ReseauLu Map showing journal publications for climate science and skeptics.

This comparative (of articles cited at least three times) shows that the climate change skeptics are indeed part of the scientific mainstream of climate change research, in the sense that they publish in top climate science journals. It also reveals that they also have their own specific outlets that publish only skeptics' research.[46] However, climate change skeptics cannot be characterized as merely a fringe based on this research. It is relevant to mention here that two separate qualitative analyses of global warming-related article abstracts through ISI have found no 'disagree[ment] with the consensus position'[47] and that 'an overwhelming percentage (97.2% based on self-ratings, 97.1% based on abstract ratings) endorses the scientific consensus on A[nthropogenic] G[lobal] W[arming]'.[48] So while the climate skeptics are part of the scientific center, this does not mean that their prominent scientific publications are by definition those in which they voice their skepticism.

46   For instance, in this sample, journals such as *Environmental Conservation*, the *Journal of GeoPhysical Research: Oceans*, and *Environmental and Experimental Botany* had only published skeptics' papers (cited at least three times and published before July of 2008).
47   N. Oreskes, 'The Scientific Consensus on Climate Change', *Science* 306.5702 (2004): 1686–1686.
48   J. Cook, D. Nuccitelli, S.A. Green, M. Richardson, B. Winkler, R. Painting, and A. Skuce, 'Quantifying the Consensus on Anthropogenic Global Warming in the Scientific Literature', *Environmental Research Letters*, 8.2 (2013): 024024.

As we have seen in the scientometric analysis of the place and status of actors within climate science (through querying ISI), selecting only cited academic papers (at least three citations) filters out the less relevant sources (i.e., those of uncited papers). The web and its search engines know a related logic that enables a means of analysis similar to citation analysis. As described by Sergey Brin and Lawrence Page in 1998, when they presented their search engine prototype, the algorithm treats hyperlinks almost like a web of science would treat a citation. 'Intuitively, pages that are well cited from many places around the web are worth looking at'.[49] [50] But not all citations are equal; those from well-cited pages have more weight. It is noteworthy that Page and Brin explicitly use the term 'citing' when they refer to linking.[51] As citations network content, scientometrics could be considered a means of networked content analysis. As scientometrics can help evaluate the weight and relevance of scientific actors and outlets, for the study of the climate controversy, it is relevant to also assess the place and status of specific actors within the broader climate change debate as it plays out on the web. This is possible with web-specific techniques of networked content analysis, as I will discuss in the following sections.

The Case of the Dutch Skeptics

On the web (broadly conceived) a national set of sources may be demarcated by taking the local domain of Google Web Search (e.g., Google.nl for the Dutch web) and querying it in the specific local (in this case Dutch) language(s). In this next section, I will zoom in on the networks and resonance of climate actors in the Dutch climate debates, to consider how networked content analysis may help to capture instances of group formation and actor resonance. I will consider moments of group formation through hyperlink networks, as I have described in the introduction; thus, rather than labeling scientists according to pre-formed categories, I understand them as part of a group when they perform as such. This approach, informed by Latour, can perhaps be best explained by example. In October 2011, the Royal Dutch Academy of Sciences (KNAW) published a report titled *Climate Change: Science and Debate*.[52] With the brochure written by a small committee of scientists from inside and outside the Academy, the KNAW set out to map the state-of-the-art of climate science, more specifically discussing what has reached scientific consensus and what still causes controversy and why. The report ends with a summary in which the topics of consensus are listed as seven statements. Statement A reads:

> Humankind changes the composition of the atmosphere quickly and drastically. The increased concentration of carbon dioxide and other greenhouse gases cannot be marginalized.[53]

49   Brin and Page, 'The Anatomy of a Large-scale Hyptertextual Web Search Engine'.
50   See also Krippendorff, *Content Analysis,* 2013, 33.
51   See also Rieder, 'What Is in a PageRank?' for a discussion of how PageRank relates to citation analysis.
52   KNAW, *Klimaatverandering, Wetenschap en Debat.*
53   KNAW, *Klimaatverandering, Wetenschap en Debat,* 34.

This first statement already is likely to turn the brochure into a controversial object, for it directly and without qualification stresses the role of humankind in global warming and the effects of CO^2 (and other emissions) on climate change. Unsurprisingly, soon after its publication, Dutch skeptical blogs started posting about this report by the KNAW, characterizing it as 'alarmist'.[54] One of the more prominent skeptical blogs of the Netherlands, climategate.nl, featured a blog posting in English stating that the brochure contained a 'tsunami of scientific errors':

> The brochure claims that these seven statements are hard science on which all scientists agree. Nothing is further from the truth: they are a rendering of the claims of the IPCC, in denial of all serious criticism that has been brought against it by the scientific community.[55]

Besides blogging about the report in various Dutch climate blogs, the skeptics chose two other formats for their criticism: a letter signed by 22 scientists demanding the retraction of the report and a *climate seminar* organized at Nieuwspoort, the international press center in The Hague.

In the letter, the scientists refute the seven statements and demand a retraction of the publication.[56] They state they represent various academic disciplines including (bio-) chemistry, physics, geology, engineering, and climatology. The only non-academic who signed the letter is Ralf Dekker, blogger and chairperson of the aforementioned Groenerekenkamer.nl. One of the scientists on the list is Pieter Ziegler, Swiss Geology Professor Emeritus at University of Basel and Emeritus Member of the Royal Academy (KNAW). For the purposes of the analysis, it is useful to consider the signatures of the letter as a ready provision or short-listing of 22 climate change skeptics. Not surprisingly, the program of the climate seminar organized by Groenerekenkamer.nl and its list of speakers was filled mostly with people on this shortlist.

The next step in the analysis of the skeptics' group formations is to study their networks, to better understand the scope and aspirations of these actors through hyperlinking, and the composition of the issue network. To generate such analyses, a list of the skeptics' websites is first entered into the IssueCrawler tool for hyperlink analysis. The IssueCrawler then performs co-link analysis, crawling the inputted (seed) 'URLs for links and retain[ing] the pages that receive at least two links from the seeds', and outputting a network graph.[57] Figure 6 (p. 55), the IssueCrawler map of Dutch skeptics, shows that the group's hyperlink network is dominated by Anglo-American sources. This is perhaps surprising given the appearance of a strong national network of Dutch skeptics with an active collective blogging culture in the Dutch lan-

---

54   T. Wolters, 'Alarmistische KNAW in Grote Problemen', 2011, http://climategate.nl/2011/10/25/alarmistische-knaw-in-grote-problemen/.

55   T. Wolters, 'Bad Science in Alarmist Report from Royal Dutch Academic Council', 2011, http://climategate.nl/2011/10/19/bad-science-in-alarmist-report-from-royal-dutch-academic-council/.

56   H. Labohm, 'Klimaatsceptici Verzoeken KNAW Klimaatrapport in te Trekken', October 2011, http://www.dagelijksestandaard.nl/2011/10/klimaatsceptici-verzoeken-knaw-klimaatrapport-in-te-trekken.

57   Govcom.org Foundation, 'IssueCrawler: Instructions of Use', http://www.govcom.org/Issuecrawler_instructions.html.

guage. The IssueCrawler map reveals, however, that these sites link not so much to each other or to other Dutch sources, but mainly to sources outside the Netherlands (see Figure 6, top).

The Dutch scientists that authored the KNAW publication (the 'non-skeptical' actors in this comparative study) show a more heterogeneous network (see Figure 6, bottom), with many Dutch sources. There is a science and government cluster in which the website of the Dutch Ministry of Foreign Affairs links to an international cluster that includes the homepages of the UN and the World Bank. There are also mainstream media clusters, involving the large Dutch daily newspapers and broadcasting companies who link to their international colleagues such as *The New York Times*, the *Financial Times*, and *La Reppublica* (Italy).

The networks immediately show two distinct actor groups. The skeptics show international aspirations in linking to their Anglo-American counterparts, and the non-skeptics reveal their rooting in science and government and their contributions to the mainstream media. To further understand their resonance within dominant sources on the topic of climate change, I proceed to use Google Web Search to select top sources and query them for the resonance of these sets of actors.

Dutch Climate Change Actor Resonance Analysis

As discussed in the previous chapter on traditions in content analysis, the demarcation of networked content is a key part of networked content analysis research, and much attention needs to be paid to the design and fine-tuning of search strings when using search engines. In this case, the demarcation of Dutch climate change sources can be operationalized by querying the Dutch Google.nl for the search term *klimaatverandering* (Dutch for 'climate change'). The top 100 results contain only 25 unique hosts consisting mainly of news sources, governmental sources, and some environmental organizations and blogs. Here, I subsequently query each of these 25 URLs for all of the 24 skeptics on the shortlist. This can be done manually, with queries such as *'Hans Labohm' site:knmi.nl* and *'Hans Labohm' site:www.wnf. nl*, and so on. At this point, I use the so-called Lippmannian device, a tool inspired by Walter Lippmann to discover partisanship, which takes as input a list of URLs and a list of queries, and then the tool does the sequencing automatically.[58] Re-sizing the URLs according to their mentioning of prior identified or short-listed skeptics then shows the sources that most involve these actors, or are most 'skeptic-friendly.' Showing a source cloud per actor and leaving the search results in their original order (i.e., of the result list in Google Web Search) renders visible that some skeptics enter into the top results, and others resonate only in the bottom results. The tool also offers a so-called 'issue cloud' in which the keywords (in this case actors' names) are clouded according to their resonance within the top sources; this shows who the most prominent actors on the shortlist are.

Figure 6 (page 55): Dutch climate actor networks. IssueCrawler maps for the Dutch climate actors (top 'skeptical,' bottom: 'non-skeptical').

---

58   See also the Digital Methods Initiative's Lippmannian Device tool page: Digital Methods Initiative, 'Lippmannian Device', https://wiki.digitalmethods.net/Dmi/ToolLippmannianDevice.

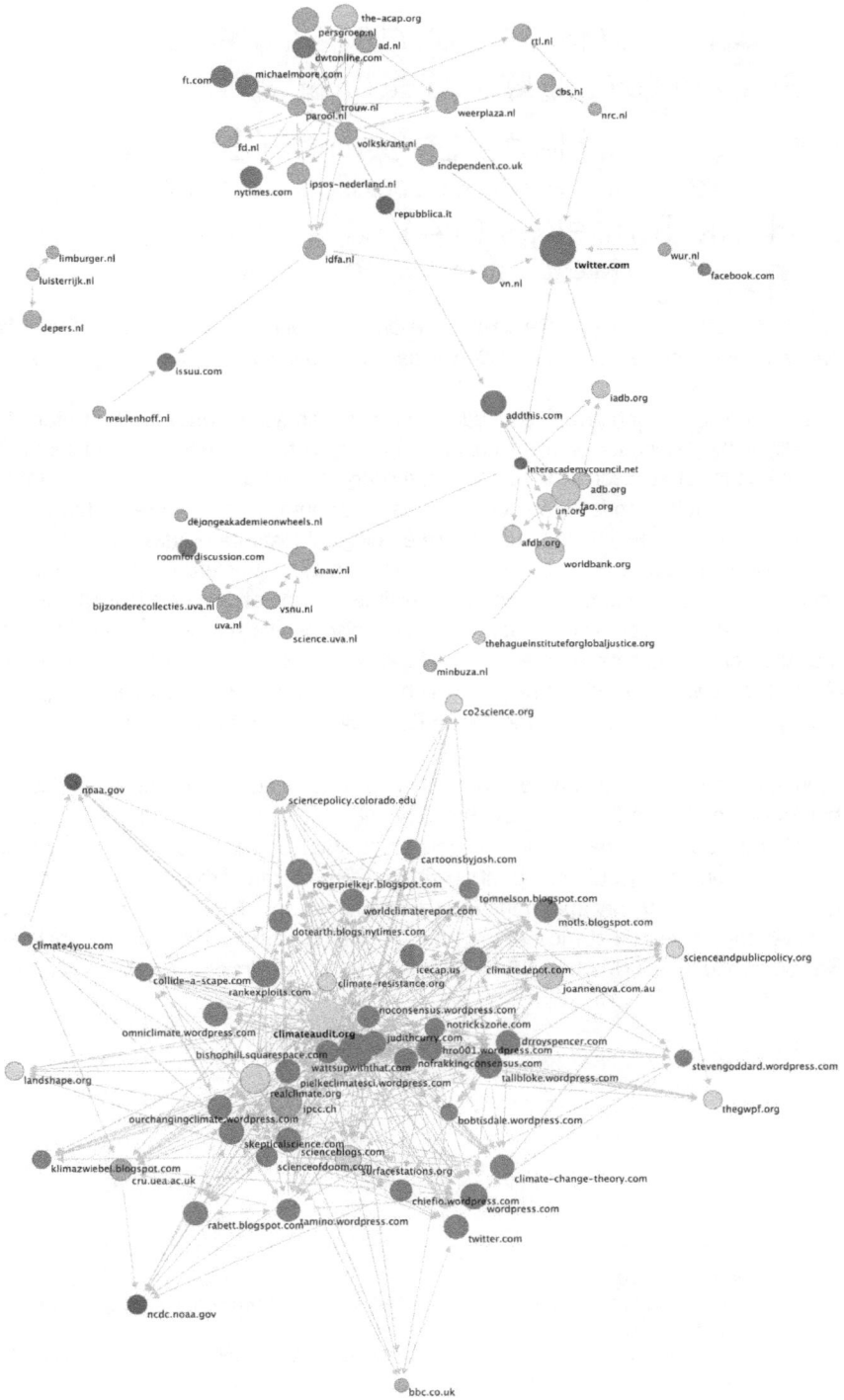

Peter Bloemers (3) Marcel Crok (10) Ralf Dekker (3) Hans Erren (3)
Bas van Geel (11) Kees de Groot (8) Albert Jacobs (2) Hub Jongen (3)

Rob Kouffeld (3) Kees Kwantes (0) Hans Labohm (15) Kees Le Pair (5)

Gerrit van der Lingen (0) Rob Meloen (3) Jan Mulderink (0) Arthur Rörsch (7) Henk Schalke (0) Hajo Smit (6) Frans Sluijter (2)
Henk Tennekes (14) Dick Thoenes (5) Theo Wolters (6)
Rypke Zeilmaker (8) Peter Ziegler (1)

Figure 7: Dutch climate change skeptics resonance cloud. Issue cloud visualizing the resonance of Dutch climate skeptics in Google search results for the query 'klimaatverandering.'

Figure 7 presents such an issue cloud (in this case, an actor cloud) for the Dutch skeptics, visualizing the resonance of the actors in the top results for the query of 'klimaatverandering'. The more the actors resonate in the results, the larger their name is depicted. The three most prominent Dutch skeptics are economist Hans Labohm, Henk Tennekes, former Director of Research at the Royal Netherlands Meteorological Institute (KNMI), and Bas van Geel, Associate Professor of Paleo-Ecology at the University of Amsterdam. Hans Labohm is an economist formerly employed by the Dutch Institute of International Relations Clingendael and, notably, a former expert reviewer at IPCC. He has also been a speaker at one of the Heartland Institute's climate skeptics conferences.[59] In 2004, Labohm published the book *Man-Made Global Warming: Unravelling a Dogma*, which he co-authored with Dick Thoenes (who is less resonant in the online debate) and Simon Rozendaal (not on the shortlist).[60]

Zooming in on Hans Labohm we can create a 'source cloud' to see which sources mention him most (see Figure 8, p. 57). Labohm generally resonates well in the media (also in *Volkskrant* and *Trouw*) and makes it into the top results. He resonates most in klimaatverandering. wordpress.com, a blog authored by atmospheric scientist Bart Verheggen, where Labohm has his own tag and category and in the Dutch daily newspaper NRC. A closer look at the NRC archives then reveals that most of this attention stemmed from 2004 when Labohm's book was published and 2007 when NRC published a portrait of him as a 'liberal' climate skeptic.[61]

---

59   ICCC4 in May of 2010.
60   H. Labohm, S. Rozendaal, and D. Thoenes, *Man-Made Global Warming: Unravelling a Dogma*, Essex: Multi-Science Publishing Co. Ltd, 2004.
61   M. aan de Brugh, 'Liberaal in de Strijd Tegen Klimaatgekte', *NRC Handelsblad*, 19 February 2007, http://vorige.nrc.nl/article1771418.ece.

knmi.nl (2) rijksoverheid.nl (2) milieucentraal.nl (0)

klimaatverandering.wordpress.com (100)
transitiontowns.nl (6) ad.nl (2) wageningenur.nl (0) duurzaamnieuws.nl (0) happynews.nl (0) scientias.nl (2) nrc.nl (100) pbl.nl (6) greenpeace.nl (1) wnf.nl (2)

volkskrant.nl (22) nu.nl (1) europa-nu.nl (7) biojournaal.nl (0) kennislink.nl (0) klimaat.startpagina.nl (0) schooltv.nl (0) encyclo.nl (14) duurzaamheidipo.kennisnet.nl (0)

standpunten.groenlinks.nl (0) trouw.nl (36)

Figure 8: Hans Labohm's source cloud. This cloud shows the resonance of Hans Labohm, the most prominent Dutch climate change skeptic, in the top results for climate change.

The sources in which skeptics resonate most are KNMI, Klimaatverandering, and NRC. There are only five sources in the results that do not mention any of the short-listed skeptics, the highest-ranked one of which is milieucentraal.nl. Milieucentraal is a foundation dedicated to providing consumers 'unbiased information on energy and environment' (Milieucentraal.nl), and its website offers hands-on tips and tricks for a sustainable or *green* lifestyle (such as reducing waste, being more energy-efficient, etc.).

Of the analyzed KNAW scientists, the author and editor of the KNAW brochure, Louise Fresco, a renowned scholar in the field of Tropical Plant Breeding and Production, Food and Agriculture, President of Wageningen University and KNAW member, is the most prominent actor (see Figure 9, p. 58). Fresco resonates in sixteen of the top climate change sources (which is only one more than Hans Labohm). In half of these sources, she is mentioned at least 100 times.[62] The second most prominent scientist is Rudy Rabbinge, Professor of Sustainable Development and Food Security at Wageningen University. The third most resonating scientist is Robbert Dijkgraaf, Director of the Institute for Advanced Study in Princeton (United States), who at the time of the publication of the report was president of the KNAW.

Collectively, the scientists resonate in all but seven of the sources. They are not present in two sources that do list skeptics, namely scientias.nl and greenpeace.nl.

In this resonance analysis, we find that there are no sources that mention only our small sample of 'non-skeptical' scientists without the short-listed skeptics. All scientists, be they climate change skeptics or not, resonate broadly in the results, both at the top and bottom of the list. So from these profiles, just as from the scientometric analysis, it is not easy to detect stark differences between the place and status of climate skeptics from that of non-skeptical scientists. But as the climate change debate takes place internationally, using networked content analysis to compare such a national *actor profile* with sets of other national actor profiles within a debate can perhaps reveal different (scientific) cultures and national frames on a global issue.

---

62   The ceiling for this scrape was set at 100, and she hits that ceiling in eight of the sources.

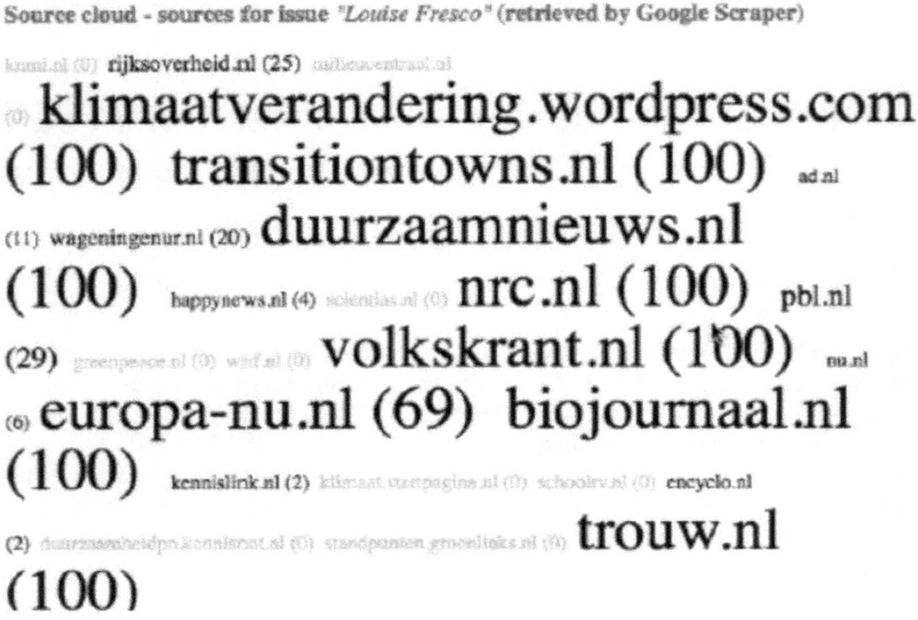

Source cloud - sources for issue *"Louise Fresco"* (retrieved by Google Scraper)

knmi.nl (0) rijksoverheid.nl (25) nslinoventraal.nl

(0) klimaatverandering.wordpress.com (100) transitiontowns.nl (100) ad.nl

(11) wageningenur.nl (20) duurzaamnieuws.nl (100) happynews.nl (4) scientias.nl (0) nrc.nl (100) pbl.nl (29) greenpeace.nl (0) wnf.nl (0) volkskrant.nl (100) nu.nl

(6) europa-nu.nl (69) biojournaal.nl (100) kennislink.nl (2) klimaat.europagina.nl (0) schoolrv.nl (0) encyclo.nl

(2) duurzaamheidpn.konsisnat.nl (0) standpunten.groenlinks.nl (0) trouw.nl (100)

Figure 9: Louise Fresco's source cloud. This cloud shows the resonance of Louise Fresco in the top results for climate change.

In a comparative analysis of these Dutch climate actors with French climate actors, I indeed noted major differences with the Dutch case and was able to demonstrate how such comparative analyses can give insight into the composition and position of these groups.[63] This French climate skepticism analysis, which I conducted together with climate journalist Denis Delbecq was already briefly discussed in the Introduction. Delbecq is an expert on the French climate debate and its prominent actors, which he appraised extensively in a 'dossier' for the French environmental journal TerraEco.[64] For this case study, Delbecq provided short lists of prominent French skeptical and non-skeptical scientists and scientific organizations. With this list, I also started by conducting hyperlink analysis, just as I did for the Dutch actors aforementioned. By linking frequently to the objects of their own criticism, the French skeptics granted high authority to these same objects, thus positioning controversy objects right in the center of their network. The IPCC was the main node in the skeptics' network. The non-skeptical scientists showed a different and much more traditional approach. These scientists granted network authority to established French scientific figures and organizations. The Dutch non-skeptical scientists also granted authority to Dutch government and media. From the resonance analysis, the most important finding was that the French skeptics, in contrast with the Dutch ones, resonated throughout the ranked results and appeared in the same outlets as their non-skeptical counterparts.

---

63   Delbecq and Niederer, 'Climatosceptiques et Climatologues'.
64   Delbecq, 'Dossier Climato-sceptiques'.

On a methodological plane, we may ask how this national perspective would be scalable to other platforms, and this is something I will assess in the next chapter (4) on Wikipedia. In ending this chapter, I want to propose the development of another actor-centric technique of networked content analysis that looks at actors within a specific issue in order to map their other issue involvements, which I show sheds further light on the actors' role and operationality in the climate change debate *as such*. As said in the introduction to this chapter, a prominent Dutch skeptic is the director of Skepsis, the skeptical organization that also addresses many other issues regarding health and religious practices. If climate change skeptics are skeptical of a range of other issues, then this arguably sheds significant extra light on my current study's consideration of these actors' operationality, opinions, and degrees of skeptical activity *within* climate change as an issue. They would be professional skeptics instead of professional climate change experts.

Do Skeptics Have Related Issues?

It is a commonplace that all issues have their skeptics, thus that the presence (and problematics) of skeptics' involvement is not at all specific to climate change. Nevertheless, as I argue that we must take the involvement (and impact) of skeptics in such a consequential debate seriously, then it is also worth asking what happens if we take the matter of climate skeptics' involvement in other issues seriously. Indeed, what would it mean to know if (climate) skeptics have *other* issues to be skeptical about? And further, how might it matter to know *which* other issues they are skeptical about?

In a previous small-scale study, I took the shortlisted climate skeptics (of the scientometric analysis) and conducted a close reading of the personal homepages. Here, I found that these prominent climate actors also publish articles and blog postings in which they present skeptical viewpoints on neighboring issues such as organic agriculture and biofuels. More unexpected — in an illuminating sense — is their skepticism on health-related issues such as the dangers of smoking and second-hand smoke, the human variety of mad cow disease (Creutzfeld-Jacob disease), and evolutionary theory.[65] In my opinion, the analysis of controversies would benefit strongly from the development of robust methods for retrieving such 'related issues,' which could be developed as part of networked content analysis. As 'climate change skeptics' are skeptical of a range of other issues, then this arguably defines them as professional skeptics. This sheds significant extra light on my current study's consideration of these actors' role *within the* climate change debate.

As we have seen in this chapter, the climate change debate, when studied only as a scientific debate (accessed through ISI), presents a scholarly space in which both non-skeptical and skeptical actors are active, publish in the same top journals (as well as separate journals for each of the groups) and have a similar distribution of scholarly disciplines. However, when addressing the same debate from a broader base, looking at the prominence and resonance of skeptical actors within climate change content on the web (accessed through Google Web Search), we are presented with distinct groups of actors and a stark profile of skeptics as

65   Niederer, 'Global Warming Is Not a Crisis!'.

professional skeptics. When the Dutch climate change publication came out in 2011, skeptics organized themselves in an event to counter the claims of consensus presented in the booklet. A closer look at the hyperlink networks of these Dutch skeptics showed their (aspired) affiliations with their Anglo-American counterparts. One of the prominent Dutch skeptics is the director of the Skepsis foundation, addressing skeptical viewpoints on a myriad of issues. This raises the question of whether these prominent skeptics are dedicated to skepticism *as such*, or to climate change as a field of knowledge production and research. Google Web Search and a close reading of the skeptics' websites gave insights into their commitments along these (divided) lines and put the scientometric analysis into a new light. It found that prominent skeptics are indeed ideologically bound, dedicated to skepticism rather than to the climate debate alone.

Conclusions

Where the Paris Agreement of 2015 marked a new phase in the climate debate, with a historic agreement but also a historically broadly perceived consensus on climate change, in this chapter I have traced back actors across science and the web, and in doing so went back in time to the first Heartland Conference of 2008. Where controversy analysis often centers on an issue (or a set of issues), the actor-centric approaches proposed in this chapter can follow actors *across* (and relevantly 'beyond') single issues as objects. This kind of analysis further complicates the characterization of climate skeptics as presented in critical literature (that, for instance, focus on industrial ties), given the revelation that these skeptics are not only focusing on climate change in their skeptical endeavors. These findings have a number of implications. First of all, on a methodological level, it provides a shift from the idea that all issues have skeptics (or that a skeptical stance is part of science) to consider the ramifications of skeptics having multiple issues. Second, as the scientometric analysis has revealed, the fact that these skeptical scientists are part of the scientific mainstream raises questions about the employment of their expertise. Why do they write about these other issues while being climate scientists? Are their publications on related issues also part of the scientific mainstream in their respective fields? Finally, we may conclude that an actor-centric approach of networked content analysis provides a means to trace a controversy and its actors outside of the boundaries of a single issue, and thus is a valuable addition to the study of actors within science (through scientometrics).

This chapter makes use of web content to research the place and status of skepticism within climate science and the climate debate. The study started with a scientometric analysis looking at the distribution of disciplines and shared places of publication of skeptical and non-skeptical actors. The scientometric data shows that climate change skeptics are part of climate science, sharing both a distribution of disciplines and a mainstream of prominent scientific outlets. Besides being sometimes at the 'center' of climate science, skeptics also work in parallel to non-skeptical climate scientists and have their respective unique journals and disciplines, their respective 'fringes' if you will.

To extend this comparison beyond academia, hyperlink analysis — in this case, in a comparison between skeptics and non-skeptics in the Netherlands — has shown the associations and aspirations of these actors. For the Dutch skeptics, this aspirational linking plays out in the prominence of skeptical Anglo-American sources appearing in their hyperlink networks. The Dutch (non-skeptical) scientists have a heterogeneous network, including science and government, as well as news media. The Dutch skeptics form an international network by linking to both (international) skeptic blogs and the subjects of their criticism. These findings here cannot be generated through citation indices and other scientometric data, but are rendered possible only through the networked content analysis techniques I have outlined.

Web resonance analysis scoring the prominence of one or more actors in a demarcated issue source sets allowed for further comparison between skeptical scientists and others. Furthermore, the output of a source cloud-enabled an analysis of actor-friendly sources. The comparative analysis reveals different 'profiles' per type of actor. The most prominent Dutch skeptics resonate well in the news and on one dedicated climate blog but as a whole resonate in fewer sources than the non-skeptical short-listed climate scientists.

Shifting focus from the issue space to an actor-centric perspective, skeptics appear to work on multiple issues, some of which are well outside of the climate science, let alone outside of climate change debates. Tools and methods like those worked through in this chapter can help to assess the commitments of individual actors to and beyond specific issues, and there-fore reveal larger stakes in a much richer and more complex ecology of related issues. Future analysis along these lines and using these methods could also benefit from a longitudinal approach, which would render visible not only the resonance of actors over time but also the top sources for the issue of climate change and their (analytical) treatment of these actors.

Where with scientometrics alone I was not able to identify the skeptics as entirely distinct from climate science, with networked content analysis I found distinct networking behavior as well as *related* issues that were objects of their skepticism (ranging from the dangers of second-hand smoke to Creutzfeld-Jacob), which qualified them as *professional skeptics* rather than professional climate experts.

Asking then what the web *does to* the climate debate, I would like to conclude that the tech-nicity of the web, with its hyperlinked websites and search engine result rankings, reveals actor-networks of affinity, association, critique (as the skeptics linking to their main object of criticism: IPCC) and aspiration, which may result in *drama* (in the case of the Dutch skeptics linking to their Anglo-American colleagues without them linking back). Search engines that rank results can be used for resonance analysis, presenting on one level the sources that make it into the top results of a query, while also offering up specific keywords or (as pre-sented in this case) actors. A close reading of these actors' websites in the presented case study of this chapter establishes a clear image of their professional skepticism, rather than a commitment to climate change as a scientific issue.

This first study sets the ground for a networked content analysis of the climate debate that is able to make use of, and also go beyond, the following of online actors and their group for-mation. The case studies that follow will apply similar novel techniques of networked content analysis to the study of the climate debate on two online platforms. As web content itself is increasingly formatted towards inclusion in such platforms, my treatment of the technicity of Wikipedia, the collaboratively authored encyclopedia project, and Twitter, the micro-blogging platform, to study the climate change debate I argue is key to comprehending the debate itself.[66] In both chapters, I will discuss the dependency of each of the respective platforms as well as their various user groups and content on the (underlying) technicity. In the case of Wikipedia, this means assessing the climate debate in this socio-technical platform for encyclopedic knowledge production, understanding the interplay between users and technical agents. In Twitter, I will address how content is networked and will further the utility of reso-nance analysis, which I deployed here in the study of climate skeptics, to see how the various stages of the climate change debate resonate. Furthermore, I will closely read clusters of hashtags in assessing the state of the climate change debate. Overall, these studies are geared towards the understanding and inclusion of technicity in the analysis of networked content.

66   Helmond, *The Web as Platform.*

4. WIKIPEDIA AS A SOCIO-TECHNICAL UTILITY FOR NETWORKED CONTENT ANALYSIS

In the previous chapter, I have assessed the climate change debate through scientometrics and networked content analysis. To understand the technicity of online networked content, I argued, it is necessary to address how content is networked and which kinds of digital methods and tools are therefore suitable for the demarcation of content and the operationalization of the research question. In this chapter, I will address the mapping of the climate debate in Wikipedia. But before coming to the discussion of this debate, as it plays out and is managed among and other controversial topics and the management thereof in Wikipedia, I will discuss how Wikipedia has revived the idea of the web as a place of human collaboration and mass participation by 'everybody'.[1] The Wikipedia platform is often considered as an example par excellence of the collaborative promise of social media, and of knowledge production and management that utilizes the wisdom of crowds. Since 2001, its group of editors and volunteers has engaged in developing an online encyclopedia whereby anyone with net access is welcome to contribute, and articles are open to continuous editing and refinement. Scholars who have evaluated or contested the value of Wikipedia content have almost unanimously focused on its crowd-based organization and have stressed the danger of producing low-quality information with many (anonymous) minds.[2]

These concerns about Wikipedia are legitimate and relevant, of course, but the one-sided focus they give to human agents while neglecting the role of technology must be both resisted and complemented by attention to the socio-technological dimension of Wikipedia as a dynamic knowledge production and management project. In this chapter, therefore, I want to explore the technicity of Wikipedia content and assess how networked content analysis can be applied to this platform. What do researchers need to know of the platform's means of content creation, networking, and maintenance to be able to analyze its content in a way that is digitally and, more specifically, platform-informed? What does Wikipedia 'do' to content that is controversial, and what does this mean for the methods of networked content analysis put forward in this book? To answer these questions, I will first analyze how dependent the human social creation, use, and maintenance of Wikipedia knowledge is upon software robots (in short referred to as bots), the non-human content agents that assist in editing Wikipedia articles. Secondly, I will discuss examples of networked content analysis of controversial content that make use of the possibilities offered up by Wikipedia's technicity for controversy research.

The technicity of Wikipedia content makes it possible to refine further the techniques of networked content analysis, and to explore how resonance, related content, actor engagement, and controversy management may be studied within this encyclopedia project. It is crucial to understand Wikipedia as a dynamic, networked encyclopedia when approaching its content for analysis, which is why I will start (as I did in the previous chapter on Twitter) with a

---

1   Shirky, *Here Comes Everybody*.
2   A. Keen, *The Cult of the Amateur: How blogs, MySpace, YouTube, and the Rest of Today's User-generated Media Are De-stroying Our Economy, Our Culture, and Our Values*, New York: Doubleday Currency, 2007.

brief introduction of the platform's technicities. Again, this is not meant to be an exhaustive overview of all the features of the platform, but rather can be seen as a kind of technical introduction to socio-technical fieldwork, exploring and describing the ways in which content is produced and networked.

Looking just at the level of its software, Wikipedia has changed drastically throughout the years. Overall, however, it remains a wiki-based encyclopedia platform, offering various levels of access to information of article history and editors, enabling researchers to follow the actors and close-read their positions, interactions, references, and commitment to a specific issue. The interface of Wikipedia presents an article and talk page for each Wikipedia subject. In the article tab, it is possible to read or edit the article, or to view the article's revision history. In the revision history, each edit is listed along with a timestamp, and a username (or IP-address for an anonymous edit). A click on the timestamp opens the particular version of the article from that edit date. It is possible to make a selection of differently dated versions of an article and compare the different revisions. For each Wikipedia article, the revision history lists external tools, including revision history statistics, revision history search, edits by user, number of watchers, and page view statistics. The Talk page shows some policies and general rules for discussion as well as a place to ask questions or discuss edits. It is also where the article's revision history is located and publicly accessible. A Wikipedia article may start with links to similarly named articles (disambiguation), or related articles. In the body text, highlighted words mark links to other Wikipedia articles. Each article ends with separate sections holding references and external links. In the left margin of the page, the language versions of the article are listed, as well as a list of 'what links here', which provides a list of all other Wikipedia articles that link to the article you have in front of you. All of this creates materials, which can be analyzed through networked content analysis.

In the next section, I will discuss how Wikipedia has been researched since its launch in 2001, and how dominant research practices have disregarded some of the crucial technical specificities of Wikipedia entailed in the production, organization, and maintenance of its content. Before discussing the climate debate in Wikipedia, I will first zoom in on two controversy analyses that are informed by the technicity of Wikipedia content, by looking at discussions on the talk pages (for the article on Gdańsk/Danzig), and by conducting a comparative analysis of articles across language versions (for the case of the Srebrenica massacre). While these analyses are unrelated to climate change research projects, they offer insights into the methodological workings of networked content analysis. The final project discussed in this chapter is a mapping of climate change articles, which ties back not only in terms of techniques but also in its subject matter to the previous chapters' case studies of the climate debate on the web accessed through Google Web Search and Twitter. In my networked content analysis here, I build on existing research to trace climate change-related content and close read actor behavior in and through Wikipedia.

Many Minds Collaborating

Wherever Wikipedia is discussed, the facts of its material composition very quickly drift into metaphor. It is variously described: by Sunstein as a platform of 'many minds' produced by what Kittur and Kraut call 'the wisdom of crowds'; by Shirky as a system of 'distributed collaboration'; by Tapscott and Williams as 'mass collaboration'; and as a space enabling hybrid new forms of 'produsage' by Bruns, inspiring what Howe calls 'crowdsourcing' and Stalder and Hirsch describe as 'Open Source Intelligence', and Poe as 'collaborative knowledge'.[3][4][5][6][7][8][9][10] As a collectively written encyclopedia launched on a wiki platform, it is indeed one of the web's most significant and longer duree (in internet history terms) examples of collaborative knowledge production. In early 2008, an article in the *New York Review of Books* explained the media cultural charm of Wikipedia:

> So there was this exhilarating sense of mission — of proving the greatness of the Internet through an unheard-of collaboration. Very smart people dropped other pursuits and spent days and weeks and sometimes years of their lives doing 'stub dumps,' writing ancillary software, categorizing and linking topics, making and remaking and smoothing out articles — without getting any recognition except for the occasional congratulatory 'barn star' on their user page and the satisfaction of secret fame. Wikipedia flourished partly because it was a shrine to altruism — a place for shy, learned people to deposit their trawls.[11]

Since the start of the Wikipedia project in 2001, the dedication of its contributors as well as the platform's success in socializing knowledge production for the benefit of many, in contradistinction to academic and media industry reliance on experts, has been through numerous waves of praise and publicly mediated criticism. While Wikipedia has indeed become famous for its collaborative approach to networks — of many minds producing knowledge — it is interesting to recall that the project originally intended to be an expert-generated encyclopedia. Beginning under the name of Nupedia, a small team of selected academics was invited to write the entries, with the aim of creating a 'free online encyclopedia of high quality'.[12] The articles would be made available to World Wide Web users through an open content license. Founder Jimmy 'Jimbo' Wales and his employee Larry Sanger put into place

3   C.R. Sunstein, *Infotopia: How Many Minds Produce Knowledge*, Oxford: Oxford University Press, 2006
4   A. Kittur and R.E. Kraut, 'Harnessing the Wisdom of Crowds in Wikipedia: Quality Through Coordination', in *Proceedings of the ACM 2008 Conference on Computer Supported Cooperative Work*, New York: ACM, 2008, pp. 37-46.
5   Surowiecki, *The Wisdom of the Crowds*.
6   Shirky, *Here Comes Everybody.*
7   D. Tapscott and A.D. Williams, *Wikinomics. How Mass Collaboration Changes Everything* (New York: Penguin, 2006).
8   A. Bruns, *Blogs, Wikipedia, Second Life, and Beyond: From Production to Produsage*, New York: Peter Lang, 2008.
9   J. Howe, 'The Rise of Crowdsourcing', *Wired Magazine* 14.6 (2006): 1–4. F.
10   Stalder and J. Hirsh, 'Open source intelligence', *First Monday* 7.6 (2002): http://firstmonday.org/htbin/cgiwrap/bin/ojs/index.php/fm/article/viewArticle/961/88. M. Poe, 'The Hive', *The Atlantic Online*, September 2006, http://www.theatlantic.com/doc/200609/wikipedia.
11   N. Baker, 'The Charm of Wikipedia,' *New York Review of Books*, 55.4 (2008), http://www.nybooks.com/articles/2008/03/20/the-charms-of-wikipedia/.
12   Shirky, *Here Comes Everybody,* 109.

a protocol based on academic peer-review.[13] [14] This expert approach failed, partly because of the slowness of the editing process by invited scholars. To speed up the process, Sanger suggested a wiki as a collective place where scholars and interested laypeople from all over the world could help with publishing and editing draft articles. The success of Wikipedia and the commitment of emerging Wikipedians took them by surprise. Sanger became the chief organizer, a wiki-friendly alternative for the job of editor-in-chief that he held for Nupedia. He made a great effort to keep Wikipedia organized while at the same time providing space for certain kinds of dynamic 'messiness' the platform was catalyzing (edit wars, inaccuracies, mistakes, fights, etc.) that ensues from collaborative production. In early 2002, however, Sanger was dissatisfied and turned away from the epistemic free-for-all of Wikipedia, towards an expert-written encyclopedic model called Citizendium; Wales stayed, choosing to pursue further the Wikipedia model.[15]

Ever since the Sanger-Wales split, the question of whether online encyclopedias and similar enterprises should be produced by a few accountable individuals (experts) or from the fruits of many (amateur) minds has been a source of heated debate. Internet critic Andrew Keen applauded Sanger for coming to his senses about the (in his view) debased value of amateur contributions in favor of professional expertise.[16] On the other end of the spectrum, many Wikipedia adepts have praised its democratizing potential as well as its ethos of community and collaborative knowledge production available to everyone to read and write.[17] [18] At the same time, the publicly consolidated narrative that Wikipedia is produced by *crowds* has been challenged, most notably by Wikipedia's founders themselves. In actuality, during the first five years of its existence, Wikipedia was largely dependent on the work of a small group of dedicated volunteers. Although they soon formed a thriving community, the notion of a massive collective of contributors was repeatedly downplayed by Wales. As he pointed out in a talk at Stanford University in 2006:

> The idea that a lot of people have of Wikipedia, is that it's some emergent phenomenon — the wisdom of mobs, swarm intelligence, that sort of thing — thousands and thousands of individual users each adding a little bit of content and out of this emerges a coherent body of work. [But Wikipedia is in fact written by] a community, a dedicated group of a few hundred volunteers. [...] I expected to find something like an 80-20 rule: 80% of the work being done by 20% of the users [...] But it's actually

---

13   Shirky, *Here Comes Everybody.*
14   Poe, 'The Hive'.
15   See also Citizendium. 'Citizendium Beta', http://en.citizendium.org/wiki/Welcome_to_Citizendium. See also historiographies of Wikipedia in: A. Dalby, *The World and Wikipedia: How We Are Editing Reality* (Somerset: Siduri Books, 2009); J.M. Reagle, *Good Faith Collaboration: The Culture of Wikipedia,* Cambridge, MA: MIT Press, 2010; and A. Lih, *The Wikipedia Revolution: How a Bunch of Nobodies Created the World's Greatest Encyclopedia,* London: Aurum Press, 2009.
16   Keen, *The Cult of the Amateur,* 186.
17   Y. Benkler, *The Wealth of Networks: How Social Production Transforms Markets and Freedom,* New Haven: Yale University Press, 2006.
18   H. Jenkins, *Convergence Culture: Where Old and New Media Collide,* Cambridge, MA: MIT Press, 2006.

much, much tighter than that: it turns out over 50% of all the edits are done by just [0].7% of the users.[19]

As Wales asserts until 2006, Wikipedia was primarily written and maintained by a small core of dedicated editors (2% doing 73.4% of all the edits). Such a disproportionate contribution of (self-)designated co-producers versus 'common users' can be found in research into production across the larger open source movement. Rishab Aiyer Ghosh and Vipul Ved Prakash were among the first to disaggregate the notion of 'many minds' collaborating in the open software movement. From their work, they conclude that 'free software development is less a bazaar of several developers involved in several projects and more a collation of projects developed single-mindedly by a large number of authors'.[20] In the open source movement then, very few total numbers of people were directly collaborating in developing software. This raises the question whether the same dynamics hold for Wikipedia.

It is important not to entirely dismiss the idea of Wikipedia's mass collectivity as a mere myth. The matter is more complicated than this. From 2004 onwards, the online encyclopedia shows a distinct decline of 'elite' users while at the same time, the number of edits made by novice users and 'masses' steadily increases. Various researchers have pointed to a dramatic shift in workloads to the common user at this point.[21] But instead of explaining the shift as a reversal of existing orders of participation, Kittur et al. speak of marked growth in the population of low edit users in terms of 'the rise of the bourgeoisie'.[22] Interestingly, these researchers explain this shift and coinage by describing Wikipedia's dynamic social system evolving as a result of the gradual development, implementation, and distribution of content management systems. After an initial period of being managed by a small group of high-powered, dedicated volunteers, the 'pioneers were dwarfed by the influx of settlers'.[23] The early adopters select and refine the technology and managerial systems, followed by a majority of novice users who begin to be the primary users of the system. Kittur and his colleagues observe a similar decline of elite users in Web 2.0 platforms and suggest that it may be a common phenomenon in the evolution of online collaborative knowledge systems.

This tentative conclusion is reinforced by the research of Burke and Kraut, which shows that to sustain the encyclopedia's growing popularity, organizers need to identify the platform's more productive workers and grant them 'administrator's status'.[24] Important to note here is that since the publication by Kittur et al. in 2007, the English-language Wikipedia has lost

---

19   A. Swartz, 'Who Writes Wikipedia', 2006, http://www.aaronsw.com/weblog/whowriteswikipedia/.

20   R.A. Ghosh and V.V. Prakash. 'Orbiten Free Software Survey', *First Monday* 5.7 (2000): 1.

21   A. Kittur, E. Chi, B.A. Pendleton, B. Suh, and T. Mytkowicz, 'Power of the Few vs. Wisdom of the Crowd: Wikipedia and the Rise of the Bourgeoisie', in *CHI*, 2007, San Jose.

22   Kittur et al. 'Power of the Few vs. Wisdom of the Crowd,' 7.

23   Kittur et al. 'Power of the Few vs. Wisdom of the Crowd,' 7.

24   M. Burke and R. Kraut, R. 'Taking Up the Mop: Identifying Future Wikipedia Administrators' in *Proceedings of the 2008 CHI Conference, Florence*, New York: ACM, 2008, pp. 3441-3446, http://portal.acm.org/citation.cfm?id=1358628.1358871.

one-third of its editors.[25][26] Problematically, the composition of this remaining editor-base mainly consists of white male editors, a gender imbalance that plays out in the substance of the encyclopedia project. 'Its entries on Pokemon and female porn stars are comprehensive, but its pages on female novelists or places in sub-Saharan Africa are sketchy'.[27]

Although Wikipedia researchers who look at compositions of the so-called crowd do observe significant historical changes in the 'wisdom of crowds' narrative, their analyses tend to retain a binary divide between (few) experts and (many) common users, without considering other factors affecting collaborative production. Where they do notice the growing presence of non-human actors, such as software tools and managerial protocols, in the evolution of Wikipedia's social dynamics, they tend to underestimate their importance. In fact, the increasing openness of Wikipedia to inexperienced human users is only made possible by a sophisticated techno-managerial system facilitating collaboration on various levels. Without the implementation of this strict hierarchical content management system and its reliance on MediaWiki software, Wikipedia would most likely have become a chaotic experiment.

According to Alexander Galloway, the Internet and many of its (open source) applications are not simply open or closed, but modulated. More specifically, Galloway's work is key to comprehending the extent to which networked technology and the management of its developments are moderated by protocol — logics and authority generated 'from technology itself and how people program it'.[28] Wikipedia, built as an open system and carried out by large numbers of contributors, appears to be a *warm, friendly* technological space, but only becomes warm and friendly through what Galloway refers to as 'technical standardization, agreement, organized implementation, broad adoption and directed participation'.[29]

It is in these formative years of Wikipedia that the specific technicity of its content materialized and developed into a techno-managerial system, imposing a hierarchical order in deciding what entries to include or exclude and what edits to allow or block.[30] Here, to look more closely at Wikipedia's organizational hierarchy (Figure 10, p. 69) is to distinguish various user groups, some of which are 'global' (in the sense that they edit across various language Wikipedias) while others are specific to a certain local Wikipedia.

25   Kittur et al. 'Power of the Few vs. Wisdom of the Crowd'.
26   Simonite, 'The Decline of Wikipedia'.
27   Simonite, 'The Decline of Wikipedia'. See also A. Halfaker, R.S. Geiger, J. Morgan, and J. Riedl,
      'The Rise and Decline of an Open Collaboration System: How Wikipedia's Reaction to Sudden
      Popularity Is Causing Its Decline', *American Behavioral Scientist* 57.5 (2013): 664–688, http://doi.
      org/10.1177/0002764212469365, for a detailed study of this problem.
28   A. Galloway, *Protocol: How Control Exists after Decentralization*, Cambridge, MA: MIT Press, 2004, 121.
29   Galloway, *Protocol*, 142.
30   Joseph Reagle has described these dilemmas and protocols around openness versus control in his
      book *Good Faith Collaboration: The Culture of Wikipedia*, in the chapter titled 'The Puzzle of Openness'
      (pp. 73–96).

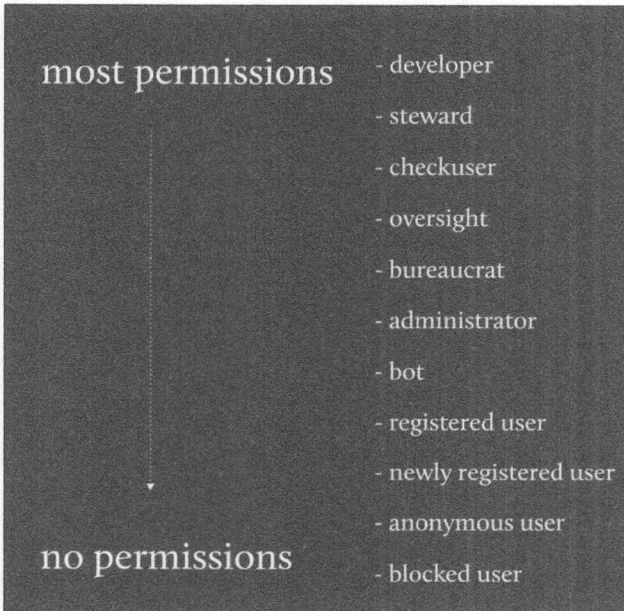

most permissions
- developer
- steward
- checkuser
- oversight
- bureaucrat
- administrator
- bot
- registered user
- newly registered user
- anonymous user

no permissions
- blocked user

Figure 10: User groups and their permission levels. Schematic overview of global and local categories of Wikipedia users according to permission levels.[31]

Each user group maintains the same pecking order, regulating the distribution of permission levels: blocked users have the least permissions, for they can only edit their own talk page. Unregistered (anonymous) users have fewer permissions than registered users, who, in turn, are at a lower level of permission than bots; bots are close to administrators (or 'admins'), who occupy the highest level in the elaborate Wikipedia-bureaucracy. System administrators (or 'developers') have the most permissions, including server access. This is a small user group of only ten people who 'manage and maintain the Wikimedia Foundation Servers'.[32] Remarkable in this ranking system is the position of bots (short for software robots), whose permission level is just below that of administrators but above the authority of registered users. I will return to the status of bots in the third section. For now, it is important just to note the significant role of automated mechanisms in the control of content.

Taking this notion of Wikipedia as a liberated vehicle of human collaboration, it could be argued that the very success of the Wikipedia project lies less as much in free collaboration as it does in the regulation of collaborative production at every level, from a small edit or a single upload to a more extensive contribution or even development of the platform or its

31 Wikimedia contributors, 'Wikipedia User Groups', 13 August 2019, http://meta.wikimedia.org/wiki/User_groups.
32 Wikimedia contributors, 'System Administrators', 3 August 2019, http://meta.wikimedia.org/wiki/System_administrators.

content.[33] Like any large public system, Wikipedia works through a system of disciplinary control by issuing rewards, such as granting a dedicated user the authority level of administrator, and by blocking the contributing rights of those users who deviate from the rules.[34] A disciplinary system of power distribution in the digital age, however, can't be regarded exclusively as a system of social control.[35] As Gilles Deleuze has pointed out in his acute revision of Foucault's disciplinary institutions, a 'society of control' deploys technology as an intricate part of its social mechanisms.[36] Wikipedia's content management system, with its distinct levels of permissions, allows moreover for protocological control: a mode of control that is at once social and technological — one cannot exist without the other.[37] Along the same lines, Bruno Latour proposes to analyze technological objects and infrastructures as 'socio-technical ensembles', in which the strict division between 'material infrastructure' and 'social superstructure' is dissolved:[38]

> Rather than asking, "is this social" or "is this technical or scientific" [...] we ask: has a human replaced a non-human? Has a non-human replaced a human? [...] Power is not a property of any of those elements [of humans or non-humans] but of a chain.[39]

Attending to the chain, rather than reinforcing the 'technology/society divide' that these theorists have already deconstructed before me, I argue that Wikipedia's dynamic interweaving of human and non-human content agents is an underrated yet crucial aspect of its performance. The online encyclopedia's success is based on socio-technical protocological control, a complex combination of its technical infrastructure and the variegated collective 'wisdom' of its contributors. Rather than assessing Wikipedia's epistemology exclusively in terms of the 'power of elites' versus the 'wisdom of crowds', I propose to define Wikipedia as a gradually evolving socio-technical system that carefully orchestrates all kinds of human and non-human contributions towards its development, by implementing managerial hierarchies, protocols, and automated editing systems that constitute the technicity of Wikipedia content. This technicity is also deployed to produce accurate and neutral content.

Accurate and Neutral Encyclopedic Information

Disregard of technological elements occurs in another heated debate haunting the Wikipedia project since its inception: the question of the credibility, accuracy, and objectivity of its con-

---

33   A lot of this is literally implemented in MediaWiki.
34   Burke and Kraut, 'Taking Up the Mop'.
35   Social scientist Mathieu O'Neil has studied the hierarchies and power structures within Wikipedia, and underlines the 'social authority' of Wikipedia administrators as 'interpreters of policy — judge, jury and executioner'. M. O'Neil, *Cyberchiefs: Autonomy and Authority in Online Tribes*, New York: Pluto Press, 2009, 159.
36   G. Deleuze, 'Society of Control', *L'autre Journal*, 1 (1990): http://www.nadir.org/nadir/archiv/netzkritik/societyofcontrol.html.
37   Galloway, *Protocol*, 17.
38   Latour, 'Technology Is Society Made Durable', 129.
39   Latour, 'Technology Is Society Made Durable', 110.

tent as an encyclopedic knowledge source, given the phenomenal difference of its experiment in socially editable, collaborated and anonymous dissemination. In other words, Wikipedia organizes the authorship of content and manages its standards, and thus 'authority', quite differently to offline projects like the Encyclopedia Britannica, against which it has often been compared and tested.[40] In response to this accuracy debate, reliant on the assumed polarity between (known) experts and (unknown) laypersons, few academics proposed to redirect their focus from encyclopedic content to the qualities and agency of Wikipedia's technological tools.

One exception is a study by historian Roy Rosenzweig that conducted a thorough analysis of Wikipedia content by comparing it biographical entries to entries from the American National Biography Online (written by known scholars).[41] Rosenzweig concludes that the value of Wikipedia should not be sought in the accuracy of its published content at one moment in time but in the dynamics of its continuous editing process — an intricate process where amateurs and experts collaborate in an extremely disciplined manner to improve entries each time they are being edited. Rosenzweig notices the benefits of multiple edits to the factuality of an entry. As he points out, it is not so many crowds of anonymous users that make Wikipedia a reliable resource, but a regulated system of consensus-based editing that shows up how history is written from multiple accounts. In his words: 'Although Wikipedia as a product is problematic as a sole source of information, the process of creating Wikipedia fosters an appreciation of the very skills that historians try to teach.'[42] One of the most important features, in this respect, is the website's built-in history page for each article, which lets you check the edit history of an entry. According to Rosenzweig, the history of an article, as well as personal watch lists and recent changes pages, are important instruments that give users additional clues to determine the quality of individual Wikipedia entries.

The politics and technicity of anonymity add a whole other layer to the accuracy debates, which is of importance to my development of networked content analysis. Disputes regarding the accuracy and neutrality of Wikipedia's content concentrate on the inherent unreliability of anonymous sources. How can an entry be neutral and objective if the encyclopedia accepts copy edits from anonymous contributors who might have a vested interest in its outcome? Critics like Keen (2007) and Denning et al. (2005) have objected to the principle of distributing editing rights to all users. What remains unsaid in this debate is that the impact of anonymous contributors is materially restricted due to technological and protocological control mechanisms. At a base level, every erroneous anonymous edit is systematically overruled by anyone who has a (similar or) higher level of permission (which is anyone except for blocked users). Since anonymous users are very low in the Wikipedia pecking order, the longevity of their edits is likely to be short when they break the rules of objectivity and neutrality. Furthermore, for anonymous editors, Wikipedia lists the IP addresses. This has inspired and enabled the creation of counter-tools such as WikiScanner for checking the identity of anonymous contributors, which it does by matching IP addresses with contact information. Bias in con-

40   See Niederer and van Dijck, 'Wisdom of the Crowd or Technicity of the Content?' for the extended
      discussion of these tests and their outcomes.
41   Rosenzweig, 'Can History Be Open Source?'.
42   Rosenzweig, 'Can History Be Open Source?', 138.

tributions can in this way be identified by a layperson, tracked across multiple entries, and if necessary, reversed.[43] My propositions for networked content analysis attendant to these socio-technics is informed by controversy mapping and follows the actors to understand the debate and the state thereof.

The debates concerning Wikipedia's accuracy and neutrality have been dominated by fallacious oppositions of human actors (experts versus amateurs, registered versus anonymous users) and have also favored a static approach to the evaluation of specific content (deemed correct or incorrect at only one particular moment in time). Both of these starting points have been ill-suited for the appreciation and analysis of dynamic and networked content in platforms such as Wikipedia, mostly because a debate grounded in such parameters fails to acknowledge the crucial impact of non-human actors—Wikipedia's dynamic content management system and the protocols by which it is run. Arguably, Wikipedia is not simply the often-advertised platform of 'many minds,' nor is it merely a free-for-all space for anonymous knowledge production. But there is more to the technicity of Wikipedia content than savvy users armed with notification feeds and monitoring devices. The technicity of Wikipedia content, key to the further development and application of networked content analysis, lies in the totality of tools and software robots used for creating, editing, and linking entries, combating vandalism, banning users, scraping and feeding content and cleaning articles. It is this complex collaboration not of crowds but of human and non-human agents combined, which defines the quality standards of Wikipedia content and is crucial to networked content analysis. These aspects must be taken into account when studying Wikipedia content.

Co-authored by Bots

The significant presence of bots in Wikipedia's workings runs counter to the commonly held assumption that Wikipedia content is authored by human crowds. In fact, human editors would be greatly strained to keep up the online encyclopedia if they weren't assisted by a large number of software robots. Bots are pieces of software or scripts that are designed to 'make automated edits without the necessity of human decision-making.' They can be recognized by a username that contains the word 'bot,' such as SieBot or TxiKiBoT.[44] Bots are created by Wikipedians, and once approved, they obtain their own user page and form their own user group with a certain level of access and administrative rights, made visible by flags on a

---

43   On the History page of each Wikipedia entry, it is possible to access the timestamp and IP-address for every anonymous edit made. The WikiScanner, a tool created by California Institute of Technology student Virgil Griffith in 2007, made it possible for anyone (not just logged in Wikipedia editors) to geo-locate anonymous edits by looking up the IP addresses in a IP-to-Geo database and listing the IP addresses and the companies and institutions they belong to, thus offering a tool for journalists trying to locate and expose biased content. In the WikiScanner FAQ on his website, Griffith states he created the WikiScanner to (among other reasons) 'create a fireworks display of public relations disasters in which everyone brings their own fireworks, and enjoys'. The WikiScanner was designed to reveal bias, and Griffith has collected the most spectacular results on his website. The Wikiscanner is now offline. On December 21, 2012, an open-source clone of WikiScanner called *WikiWatchdog* was launched. F. Scrinzi and P. Massa, 'WikiWatchDog', 2010, http://www.wikiwatchdog.com.

44   The name 'bot,' and my description here of their movements may make bots appear as elaborate kinds of Artificial Intelligence robots but in fact they are mostly very simple scripts that are triggered by rules.

user account page. One year after Wikipedia was founded, bots were introduced to help with repetitive administrative tasks. Since the first bot was created on Wikipedia, the number of bots has grown exponentially. In 2002, there was only one active bot on Wikipedia; in 2006, the number had grown to 151, and in 2008 there were 457 active bots.[45][46]

In general, there are two types of bots: editing (or 'co-authoring') bots and non-editing (or 'administrative') bots. Each bot has a very specific approach to Wikipedia content, related to its often-narrow task. Administrative bots are most well known and well-liked among Wikipedia users, deployed to perform policing tasks, such as blocking spam and detecting vandalism. Bots that combat vandalism come into action when seemingly radical or destructive edits are made, for example, when large sections of content are deleted or written over in an article. Spellchecking bots check language usage and make corrections in Wikipedia articles. Ban enforcement bots can block a user from Wikipedia, and thus take away his or her editing rights, which is something a registered user is not able to do. Non-editing bots also include data miners, used to extract information from Wikipedia, and copyright violation identifiers. The latter compare text in new Wikipedia entries to what is already available on the web about that specific topic and report this to a page for human editors to review. Most bots are created to perform repetitive tasks and make many edits. In 2004, the first bots had accrued a record number of 100,000 edits.

The second category of editing or co-authoring bots seems to be much less known by Wikipedia users and researchers (for otherwise, they would certainly have played a role in the debates about reliability and accuracy). While not every bot is an author, all bots can be classified as what I am calling *content agents*, as they all actively engage with Wikipedia content. The most active Wikipedians are, in fact, bots; a closer look at various user groups reveals that bots create a large number of revisions with high quality. Adler et al. discovered that the two top contributors in their test of the longevity of edits were bots.[47] As mentioned before, bots as a user group have more rights than registered human users and also a particular set of permissions. For instance, bot edits are by default invisible in recent changes logs and watch lists. Research cited above has already pointed out that Wikipedians rely on these notification systems and feeds for the upkeep of articles.

Describing Wikipedians in bipolar categories of humans and non-humans, however, does not do justice to what is the third category of many active users being robustly assisted by administrative and monitoring tools. The capacities of these kinds of users are captured in naming them 'software-assisted human editors.' Bots are Wikipedians' co-authors of many entries. One of the first editing bots to be deployed by Wikipedians was rambot, a piece of software created by Derek Ramsey.[48]

45 Wikimedia contributors, 'Bot Activity Matrix', http://stats.wikimedia.org/EN/BotActivityMatrix.htm.
46 Wikimedia contributors. 'Editing Frequency of All Bots', 3 March 2018, http://en.wikipedia.org/wiki/Wikipedia:Editing_frequency/All_bots.
47 Adler et al, 'Measuring Author Contributions to Wikipedia', in *Proceedings of WikiSym 2008*, Porto, New York: ACM, 2008, https://users.soe.ucsc.edu/~luca/papers/08/wikisym08-users.pdf.
48 Wikimedia contributors, 'User:Ram-Man', 1 March 2016, https://en.wikipedia.org/w/index.php?title=User:Ram-Man&oldid=707772255.

La Grange, Illinois

From Wikipedia, the free encyclopedia

This is an old revision of this page, as edited by Rambot (talk | contribs) at 15:32, 11 December 2002 (misc updates). The present address (URL) is a permanent link to this revision, which may differ significantly from the current revision.

(diff) ← Previous revision | Latest revision (diff) | Newer revision → (diff)

La Grange is a village located in Cook County, Illinois. As of the 2000 census, the village had a total population of 15,608.

Geography

La Grange is located at 41°48'29" North, 87°52'24" West (41.807938, -87.873455)[1].

According to the United States Census Bureau, the village has a total area of 6.5 km² (2.5 mi²). 6.5 km² (2.5 mi²) of it is land and none of it is covered by water.

Demographics

As of the census of 2000, there are 15,608 people, 5,624 households, and 4,049 families residing in the village. The population density is 2,400.9/km² (6,220.7/mi²). There are 5,781 housing units at an average density of 889.3/km² (2,304.1/mi²). The racial makeup of the village is 91.02% White, 6.02% African American, 0.09% Native American, 1.00% Asian, 0.02% Pacific Islander, 0.99% from other races, and 0.86% from two or more races. 3.66% of the population are Hispanic or Latino of any race.

There are 5,624 households out of which 37.9% have children under the age of 18 living with them, 60.3% are married couples living together, 9.2% have a female householder with no husband present, and 28.0% are non-families. 24.5% of all households are made up of individuals and 9.7% have someone living alone who is 65 years of age or older. The average household size is 2.67 and the average family size is 3.23.

In the village the population is spread out with 28.5% under the age of 18, 4.8% from 18 to 24, 29.4% from 25 to 44, 23.8% from 45 to 64, and 13.5% who are 65 years of age or older. The median age is 38 years. For every 100 females there are 94.2 males. For every 100 females age 18 and over, there are 87.4 males.

The median income for a household in the village is $80,342, and the median income for a family is $95,554. Males have a median income of $62,030 versus $41,260 for females. The per capita income for the village is $34,887. 4.0% of the population and 3.2% of families are below the poverty line. Out of the total people living in poverty, 4.3% are under the age of 18 and 4.4% are 65 or older.

La Grange, Illinois

From Wikipedia, the free encyclopedia   Coordinates: 41°48'29"N 87°52'24"W

This is an old revision of this page, as edited by Gilliam (talk | contribs) at 01:59, 14 November 2015 (Reverted good faith edits by 107.135.57.3 (talk): Citation needed. (TW)). The present address (URL) is a permanent link to this revision, which may differ significantly from the current revision.

(diff) ← Previous revision | Latest revision (diff) | Newer revision → (diff)

For the unincorporated community in Brown County, see La Grange, Brown County, Illinois.

La Grange, a suburb of Chicago, is a village in Cook County, in the U.S. state of Illinois.[1] The population was 15,550 at the 2010 census.[2]

Contents [hide]

La Grange, Illinois
Village

La Grange Village Hall

Country	United States
State	Illinois
County	Cook
Coordinates	41°48'29"N 87°52'24"W
Area	2.52 sq mi (7 km²)
- land	2.52 sq mi (7 km²)
- water	0.00 sq mi (0 km²)
Population	15,550 (2010)
Density	6,170.6 / sq mi (2,382 / km²)
Timezone	CST (UTC-6)
- summer (DST)	CDT (UTC-5)
Postal code	60525
Area code	708

History

This section **does not cite any sources**. Please help improve this section by adding citations to reliable sources. Unsourced material may be challenged and removed. (April 2012)

The area around La Grange was first settled in the 1830s, when Chicago residents moved out to the west due to the rapid population increase in the city in the decade since its incorporation. The first settler, Robert Leitch, came to the area in 1830, seven years before the City of Chicago was incorporated. La Grange's location, at approximately 13 miles (21 km) from the Chicago Loop, is not considered far from the city by today's standards, but in that time the residents enjoyed the peace of rural life without much communication with urban residents.

The village was officially incorporated on June 11, 1879. It was founded by Franklin Dwight Cossitt, who was born in Granby, Connecticut, and raised in Tennessee, and moved to Chicago in 1862 where he built a successful wholesale grocery business.

Figure 11: A bot-created article compared to a human-edited article. The top screenshot is the La Grange, Illinois article as created by rambot on 11 December 2002. The bottom screenshot shows the same article on 14 November 2015.[49]

49   Wikipedia contributors, 'La Grange, Illinois', 27 February 2016, https://en.wikipedia.org/w/index.

Rambot pulls content from public databases and feeds it into Wikipedia, creating or editing articles on specific content, either one by one or as a batch. Since its inception in 2002, rambot has created approximately 30,000 articles on U.S. cities and counties on Wikipedia using data from the CIA World Factbook and the U.S. Census. Since the content produced by authoring bots relies heavily on their source, errors in the data set caused rambot to publish around 2,000 corrupted articles. With time, bot-generated articles on American cities and counties were corrected and complemented by human editors, following a strict format protocol: history, geography, demographics, etc. The articles appear strikingly tidy and informative and remarkably uniform. If we compare, for instance, an article on La Grange, Illinois, as created by rambot in 2002 with a more recent version of this article from 2009, it clearly shows the outcomes of a collaborative editing process; the entry has been enriched with facts, figures and images (Figure 11, p. 74). The basic format, however, has remained the same. To date, it still is rambot's main task to create and edit articles about US counties and cities, while human editors check and compliment the facts provided by this software robot.[50]

But how dependent is Wikipedia on the use of bots as content agents for the creation and editing of its articles? What is the relative balance of human versus non-human contributions in the online encyclopedia? Peculiarly, the answer to this simple question turns out to be layered and nuanced. From the statistics offered by Wikipedia, it is observable that the use of non-human contributions differs to a striking degree between various language Wikipedias.[51] As a global project, Wikipedia features over ten million articles in over 250 languages. What is the relative balance of human versus non-human agents? The fact that Wikipedia distinguishes between local and global user groups already suggests that bot activity might differ across local Wikipedias. As it turns out, specific language Wikipedias not only greatly vary in size and number of articles, but also in bot activity. The percentage of bot edits in all Wikipedias combined was 21,5% in 2009. In 2014, Wikipedia had 22.4% bot activity. The percentage of bot edits in all Wikipedias combined was 25,8% in February of 2015. Excluding the English language Wikipedia, total bot activity counts up to over 35% (which was 39% in 2009). This shows that bot activity is unevenly distributed across language versions.[52]

To account for the differences in bot activity versus human activity, in previous research I have compared bot activity in the most-used language Wikipedias (English, Japanese, German) to bot activity in endangered and revived language Wikipedias (e.g., Cornish, Oriya, Ladino).[53] Most of the editing of the English, Japanese, and German Wikipedias in 2008 was shown

php?title=La_Grange,_Illinois&oldid=707244890.

50   See SmackBot's request for approval here: http://en.wikipedia.org/wiki/Wikipedia:Bots/Requests_for_
     approval/SmackBot_0.

51   'Wikimedia Statistics', http://stats.wikimedia.org/.

52   See also 'Wikimedia Statistics', http://stats.wikimedia.org/.
     Researchers have also studied controversial 'forkings' (or splitting) of language versions, most famously
     the Spanish fork of 2002, a full copy of the Spanish Wikipedia content to a new wiki with the name
     'Enciclopedia Libre,' which left the 'Spanish Wikipedia rather inactive for all of 2002'. Lih, The Wikipedia
     Revolution, 138.
     See also: N. Tkacz, Wikipedia and the Politics of Openness, Chicago: University of Chicago Press, 2015.

53   Digital Methods Initiative, 'Networked Content', 2008, https://digitalmethods.net/Digitalmethods/
     TheNetworkedContent.

to be done by human editors. The German Wikipedia, for instance, had only 9% bot activity, the English version even less. Wikipedias of small and endangered languages showed a high dependency on bots and a relatively small percentage of human edits. One small Wikipedia, in the language 'Bishnupriya Manipuri' had seen 97% of its edits made by bots. Further analysis of bot activity versus human activity revealed that the degree of bot dependency could be an indicator of the general state of a language Wikipedia — if not the state of that language itself — in the global constellation.

It is noticeable when looking at the different types of bots that Wikipedias are maintained mainly by bots that network the content. These are called interwiki and interlanguage bots. These bots take care of linking 'articles to articles' in Wikipedias, and prevent links and pages from becoming orphans or dead ends. Wikipedia policy states that all articles should be networked and part of the Wikipedia web. Not only are 'good' Wikipedia articles full of links to reliable sources, but they should also link to related Wikipedia articles and sub-articles, and be linked to. Articles that only refer to each other, but are not linked to or linking to other articles, are also considered a threat to the principle of building the web.[54] Most of the work in interlinking these Wikipedia language versions is done by so-called interwiki bots.

It is possible to analyze a language version's state of interconnectedness using the Wikipedia statistics pages, featuring lists of the most active bots per language Wikipedia. They reveal that most-used language Wikipedias, which obviously contain much more content than the smaller language Wikipedias, have bot activity distributed across administrative tasks. In German, for instance, the top 45 of most active bots featured 27 interwiki bots and 18 bots that are meant to edit content, add categories and fix broken links. In the smaller language Wikipedias, bots significantly outnumbered human editors and were mostly dedicated to linking articles to related articles in other Wikipedias; they made sure the content, however scarce, is networked. The Cornish Wikipedia's top 45 of most active bots, for instance, showed at least 35 interwiki bots, and the remainder were bots with unspecified functions. These interwiki bots, such as Silvonenbot, a bot that adds interlanguage links, make connections between various language Wikipedias. Smaller language Wikipedias thus make sure that every article is properly linked sideways and prevent the language Wikipedia from becoming isolated.

Tracing the collaboration between human and non-human agents in Wikipedias thus allows for interesting and unexpected insights into the culturally and linguistically diverse makeup of this global project. Following the 'wisdom of crowds' paradigm, it is tempting to look for cultural-linguistic diversity in patterns of transnational collaboration in different languages, from so many proliferated cultural backgrounds. But in line with this paradigm, British information scientists have demonstrated that the Internet – and Wikipedia in particular – is anything but a culturally neutral space; major aspects of online collaborative work are influenced by pre-existing cultural differences between human contributors, as discussed in a comparative content analysis of the editing behavior found in four language versions of the Wikipedia article

54   See also Wikimedia contributors, 'Wikipedia:Manual of Style/Linking', 5 March 2016, https://en.wikipedia.org/w/index.php?title=Wikipedia:Manual_of_Style/Linking&oldid=708334675.

on Games.[55] Adding a medium-specific networked content analysis of the varied distributions of bot dependency across the wide range of language Wikipedias, it is possible to elaborate further that cultural differences in collaborative authoring of Wikipedia content cannot just be accounted for in terms of human users; they reveal themselves, perhaps more strikingly, in the relative shares of human and non-humans contributions, which can be tracked through automated patterns of contributions. High levels of bot activity, mainly dedicated to networking content and to building the web, are an indicator of small or endangered languages; a wider variety of bot activity, largely subservient to human editing activity, could be considered an indicator of a large and lively language space. This is relevant to the understanding of Wikipedia content, for those researchers invested in its analysis.

Before moving to the climate debate, in the following section, I will present two studies that each offer a close reading of articles in order to study a controversy (in this case the Srebrenica Massacre and the city name of Gdańsk) and how it is taking place behind the scenes of Wikipedia articles. I discuss these studies to make a case for an approach to networked content analysis that uses the (ever-evolving) technicity of the Wikipedia platform in the analysis of a controversial topic. Subsequently, I will proceed to discuss the issue central to the book, namely that of the climate change debate. The study explicitly deploys the *networked-ness* of Wikipedia content to demarcate an arrangement of related, interlinked articles and looks into the composition of its editors as well as editing activity over time.

Wikipedia and Controversy Mapping

In its status as an encyclopedia project, it seems initially counterintuitive to think of Wikipedia as a space of controversy. If it were to operate fully in line with the offline genre of the encyclopedia, as a utility whose information is pre-officiated and fixed (but indeed, revisited authoritatively with each edition) the online reader would assume that all controversy would aim to be resolved as best as possible, prior to its publication. However, due to the way Wikipedia content is networked, designed, and managed, the platform has emerged to be recognized as a unique socio-technical site of, and for, controversy mapping, an encyclopedic project that is ever exposed 'in the making'. To deal with controversy at the level of information, Jimbo Wales advocates the description of sometimes-conflicting perspectives within the same article, to achieve a neutral point of view (the NPoV rule). In his words:

Perhaps the easiest way to make your writing more encyclopedic is to write about what people believe, rather than what is so. In making this work, the NPoV rule in Wikipedia is crucial and has therefore been heralded as a success story of the potential of open editing. Consider the example of the controversial entry on abortion, where, after a dispute, editors chose to include an in-depth discussion of the different positions about the moral and legal viability of abortion at different times. [...] This made it easier to organize and understand the arguments surrounding the topic

---

55   U. Pfeil, P. Zaphiris, C.S. Ang, 'Cultural Differences in Collaborative Authoring of Wikipedia', *Journal of Computer-Mediated Communication* 12.1 (2006): 88–113.

of abortion, which were each then presented sympathetically, each with its strengths and weaknesses.[56]

There are other examples in which a networked content analysis of controversial Wikipedia articles provides a much richer view of the debates taking place around a particular topic than the site itself can achieve. For instance, using the different language versions of an article is a useful means to compare Wikipedia articles on a single specific issue. Researchers including Rogers and Sendijarevic, and similarly Bilic and Bulian, have pointed out that it is more accurate to say that there are 'national' rather than 'neutral' points of view, where different language versions provide different views on a specific historic event.[57] In this section, I will discuss two analyses of controversy around the history of a specific place, and how these case studies deploy the technicities of Wikipedia content for their analysis. First, I will discuss a famously debated article on Gdańsk/Danzig.[58] Secondly, I will discuss the study of the Srebrenica massacre by Rogers and Sendijarevic.

The article on Gdańsk/Danzig is one of the better-known controversy objects within Wikipedia.[59] An ethnographic study by Darius Jemelniak explores this case extensively, by looking at how the 'traditional dispute resolution methods' of Wikipedia proved ineffective in this case, such that consensus was never reached.[60] The article on Gdańsk, which was written already in 2001 with the start of the Wikipedia project, in its first version consisted of just two sentences: 'Gdańsk is a city in Poland, on the Baltic sea. Its old German name is Danzig.'[61] In December of the same year, after several changes to the body of the article, an editor decided to change its title and all other mentions of Gdańsk in the article to Danzig. Jemelniak describes how various editors have striven to reach a compromise in both the naming and the description of the city and its history through traditional means of conflict resolution, such as discussion on the talk page, mediation by administrators in contributing to the article, closing down the article from editing activity and eventually splitting the article into one about Gdańsk and one about Danzig.

Jemelniak emphasizes that in accordance with the larger Wikipedia model, a consensus is often reached over time; therefore, 'winning an argument is simply about staying in the discussion long enough'.[62] In the case of Gdańsk, however, longevity did not lead to consensus,

---

56   Wales in Bruns, *Gatewatching*, 112.
57   R. Rogers and E. Sendijarevic, 'Neutral or National Point of View? A Comparison of Screbrenica Articles across Wikipedia's Language Versions', presented at the Wikipedia Academy 2012, Berlin, 2012.
58   D. Jemelniak, *Common Knowledge? An Ethnography of Wikipedia*, Stanford, CA: Stanford University Press, 2014.
59   Wikipedia contributors, 'Gdańsk', 10 March 2016, https://en.wikipedia.org/w/index. php?title=Gda%C5%84sk&oldid=709411660.
60   Jemelniak, *Common Knowledge?*, 59.
61   Jemelniak, *Common Knowledge?*, 65.
62   In his book chapter on the controversy, Jemelniak describes the various editor types that remained active throughout the years and distinguishes between 'at least four groups', including German and Prussian nationalists (pro-Danzig), Polish nationalists (pro-Gdańsk), editors trying to end the dispute by looking at sources (no preference), and editors trying to end the dispute through mitigation and inclusion of all viewpoints. Jemelniak, *Common Knowledge?*, 67.

and the edit war persisted for years. Between 2003 and 2005, the editing was mainly done by four editors heatedly working on the article, which lead administrator Ed Poor (who we will see more of in the study of the climate change articles) to intervene. His efforts however, only exacerbated the edit war, which was by then even listed as one of the 'lamest edit wars' ever on the Wikipedia page dedicated to tracking these.[63]

Eventually, a sub-page was set up for voting about the naming convention. This subpage first 'prolonged the debate' but later did facilitate a vote, which attracted a strikingly small number of only 80 votes.[64] Today, the Gdańsk page in English uses the city name Gdańsk throughout the article, and Danzig has its own dedicated article. Where Jemielniak looks mostly at the various actors and their discussions in the talk page for his content analysis, which allows for a close reading of the controversy, he also makes use of the technicity of the platform that includes the editing history per user, and checks the editing history of some of the 80 editors who did vote, for instance. The fact that some of these editors only had a very limited editing history before the date of the vote raises further questions about whether user accounts were created solely for this purpose.[65] Jemielniak's analysis concludes from this that Wikipedia as a 'community relies as much on cooperation as it does on conflict', which he then fleshes out by looking at the strict editing protocols at play (discussed earlier in this chapter).[66]

In his analysis, Jemielniak makes use of various technicities of Wikipedia. For instance, he looks at the history of the article comparing versions of the article, follows the debate on the talk page, studies the actor composition by looking at the different users in the editing history, and looks at editing activity per user and the profiles of each of the Wikipedians involved in the discussions and editing wars. Furthermore, he gains insight into the internal Wikipedia culture by describing the role of administrators in mediating and locking down controversial articles, and by pointing at the (humorously intended) 'lamest edit wars' page.[67] However, where Jemelniak starts his study by saying that 'traditional dispute resolution methods' did

---

63   Poor suggested the following solution: 'Gdańsk (or Danzig) is a famous European city with a long and colourful history. It is known in English by two slightly different names: in alphabetical order, *Danzig* (German) and *Gdańsk* (Polish)'. Jemielniak, *Common Knowledge?*, 69.
      See also: Wikimedia contributors, 'Lamest Edit Wars', 17 July 2019, http://en.wikipedia.org/wiki/Wikipedia:Lamest_edit_wars.

64   Jemielniak, *Common Knowledge?*, 73.

65   The first sentence of the Gdańsk article reads: 'Gdańsk (pronounced [gdansk], English pronunciation gdænsk/, German: Danzig, pronounced [ndantsnç], also known by other alternative names) is a Polish city on the Baltic coast, the capital of the Pomeranian Voivodeship, Poland's principal seaport and the centre of the country's fourth-largest metropolitan area.' Wikipedia contributors, 'Gdańsk', 10 March 2016, https://en.wikipedia.org/w/index.php?title=Gda%C5%84sk&oldid=709411660.
      The first sentence of the 'Free City of Danzig' article reads 'The Free City of Danzig (German: Freie Stadt Danzig; Polish: Wolne Miasto Gdańsk) was a semi-autonomous city-state that existed between 1920 and 1939, consisting of the Baltic Sea port of Danzig (now Gdańsk, Poland) and nearly 200 towns in the surrounding areas.' Wikipedia contributors, 'Free City of Danzig', 14 August 2019, http://en.wikipedia.org/wiki/Free_City_of_Danzig.

66   Jemielniak, *Common Knowledge?*, 84.

67   The questions of which lock-down mechanisms are deployed by Wikipedia and what is the role of bots (and their automated user blocking) in these edit wars are worth asking here too.

not work in the Gdańsk/Danzig example, we will see that the eventual forking of the article (into one about Gdańsk and one about Danzig) to displace controversy is a means to end (or at least isolate) controversy. This strategy is used frequently in Wikipedia, and may even be one of the most relied upon, and appreciated dispute resolution mechanisms.

Another strong example of a study that makes use of the technicity of Wikipedia content to appraise controversy in the workings of Wikipedia was conducted by Rogers and Sendijar-evic around the topic of the Srebrenica massacre of July 1995. Where Jemielniak describes Wikipedia as a dissent-driven platform, Rogers and Sendijarevic discuss the platform as a 'cultural reference', and site for controversy mapping.[68] Perhaps it is needless to emphasize again that this is a counter-intuitive point of departure from the notion of Wikipedia authorship as being principally invested in the cultivation of a neutral point of view (NPoV), to '[represent] fairly, proportionately, and as far as possible without bias, all significant views that have been published by reliable sources'.[69] In this case study, conducted by Rogers and Sendijarevic, the research question is whether Wikipedia could show up ongoing differences in points of view on the events of July 1995 in Srebrenica, through a method of comparing various language versions of the article on the Srebrenica Massacre.

The analysis compares six language versions of the article on the 'Srebrenica Massacre', namely the English, Dutch, Bosnian, Serbian, Croatian, and Serbo-Croatian versions. The content used for comparison contains the common parts of an article, such as title, table of content, authors (or editors), images, and references. Wikipedia-specific content elements that are added to the data set include the discussion pages and the location of anonymous editors (based on their IP-address). This leaves out other similarly specific elements that are also of interest in the study of Wikipedia articles, such as the activity of bots, which as discussed, are often the most active editors, whether across an entire language version of Wikipedia or in a single article.

A first step in the analysis was to align side-by-side the different elements of the various arti-cles. Tables and charts were drawn up, which enabled the researchers to quickly discover that, indeed, significant discrepancies between the different language versions could be discerned. First of all, in the article titles: 'Srebrenica Massacre' (English), 'Masakr u Srebrenici' (Serbi-an), 'Masakr u Srebrenici' (Serbo-Croatian), 'Genocida u Srebrenici' (Bosnian), 'Genocide u Srebrenici'(Croatian) and 'De Val van Srebrenica' (Dutch), they could identify references to this single event as massacre, genocide, or the military term 'fall' of Srebrenica, as the Dutch article title reads. Another striking difference could be found in the victim count across article versions (Table 1, p. 82), where the Dutch and Serbian articles round down, and the others tend to be higher, and the English one most specific.[70]

---

68   Rogers and Sendijarevic. 'Neutral or National Point of View?'
69   'Wikipedia: Neutral Point of View', 2012, https://en.wikiepdia.org/wiki/Wikipedia:Neutral_point_of_view.
70   Rogers and Sendijarevic. 'Neutral or National Point of View?'

Wikipedia Language version	Number of Bosniak victims of the Srebrenica massacre
Dutch (Nederlands)	7000-8000
English	8372
Bosnian (Bosanski)	8000
Croatian (Hrvatski)	8000
Serbian (Srpski)	6000-8000
Serbo-Croatian (Srpsko-Hrvatski)	8000

Table 1: Wikipedia articles compared across language versions. Comparison of victim counts from the Srebrenica massacre in the Bosnian, Croatian, Dutch, English, Serbian and Serbo-Croatian articles.[71]

The first analysis confirmed a 'national' point of view rather than a 'neutral' point of view.[72] With methodological nuance, Rogers and Sendijarevic explore Networked Content Analysis on different technical levels. Firstly, on a Wikipedia language version level, their detailed findings give an overview of the four Balkan language versions (Serbian, Croatian, Serbo-Croatian and Bosnian), and compare them in terms of article count, number of edits, number of users and number of active users. Secondly, on this same level, they compare the creation dates of the various Srebrenica massacre articles in the respective Wikipedia language versions, including Dutch and English, and set these against the creation dates of the Wikipedia language versions themselves.

Analyzing the editors of these articles for each language version, Rogers and Sendijarevic's results show editor activity across language versions, and for the anonymous users (for which an IP-address is listed as mentioned before when discussing the WikiScanner) an overview of their location. (Interestingly, as Networked Content Analysis researchers you can localize anonymous users, but not registered ones.) At the level of the article, their study includes a comparison of the use of images 'looking at the sheer numbers (62 in total), the shares of them (English with 20, Bosnian 15, Croatian 14, Serbian and Serbo-Croatian 5 and Dutch 3), the common ones, and those that are unique', and a similar analysis of shared and unique references, the victim count per article, and the table of content.[73] Regarding the talk

71   Rogers and Sendijarevic. 'Neutral or National Point of View?'.
72   Rogers and Sendijarevic. 'Neutral or National Point of View?'
73   Rogers and Sendijarevic. 'Neutral or National Point of View?', 46.

page, their study offers a very detailed description of the actors' positions and discussions. Rogers and Sendijarevic make the point that these sub-analyses, especially of discussions, show the struggles to achieve neutrality, especially in the English and Serbo-Croatian version. 'Editors of the various language versions participate in the English version, which results in a continually contested article often referred to (in the Serbian article) as western biased. The Serbo-Croatian strives to be anti-nationalist and apolitical, employing a variety of means to unify the Bosnian and Serbian points of view.'[74] In all, the researchers found that 'the analysis provides footing for studying Wikipedia's language versions as cultural references'.

Both the Danzig and Srebrenica study offer examples of how the technicity of Wikipedia content provides opportunities for controversy mapping. A good example of what a Networked Content Analysis approach could look like when applied to the issue of climate change on Wikipedia can be found in a study by digital methods researchers Carolin Gerlitz and Michael Stevenson, which was conducted already in 2009, and is discussed in the following section. Their case study, titled *The Place of Issues*, combines the study of networked articles with a close reading of editing activity, and actor commitment, including active bots.[75]

Wikipedia and the Climate Change Debate

In their study, Gerlitz and Stevenson first collect all Wikipedia articles that are interlinked with the article on Global Warming, and only retain the reciprocal links.[76] Subsequently, each of the resulting URLs is scraped for links to Wikipedia articles, which are collected in a relational database. This database is visualized with ReseauLu, software for network analysis and visualization, after which the articles selected for further analysis are highlighted (Figure 12, p. 83).

The technicity of this Wikipedia article ecology represents a historical and geographical 'mapping' of a dispute that can be studied through a Networked Content Analysis. The network graph displays the network of articles surrounding 'Global Warming' on Wikipedia, based on links between the articles. The nodes are sized according to their numbers of links, and shaped according to their role in the network (hubs appear in purple), and distributed according to the links they receive (in-degree centrality) and give (out-degree centrality) to other articles. The article 'Global Warming' acts as a central node, connecting a dense cluster of articles related to climate change science (e.g. temperature records, key reports and concepts), to a looser, more heterogeneous network of articles, including some of the terms most popularly associated with the issue ('Climate Change', 'Carbon Dioxide', 'Ozone Depletion', 'Kyoto Protocol' and 'Renewable Energy'). Notably, this last group includes articles explicitly about the climate change debate: e.g., 'Scientific Opinion on Climate Change', 'Global Warming Controversy', and 'Solar Variation' (considered by the Wikipedian who created the article as 'competition for "global warming" theory').[77] Within both clusters are articles explicitly

---

74   Rogers and Sendijarevic. 'Neutral or National Point of View?', 1.
75   Digital Methods Initiative, 'The Place of Issues', 2009, https://wiki.digitalmethods.net/Dmi/
     ThePlaceOfIssues.
76   Digital Methods Initiative, 'The Place of Issues'.
77   Digital Methods Initiative, 'The Place of Issues'.

about climate change debates, such as 'Scientific opinion on climate change' and 'Global warming controversy' in the looser cluster, and 'Climate change denial' in the dense cluster.

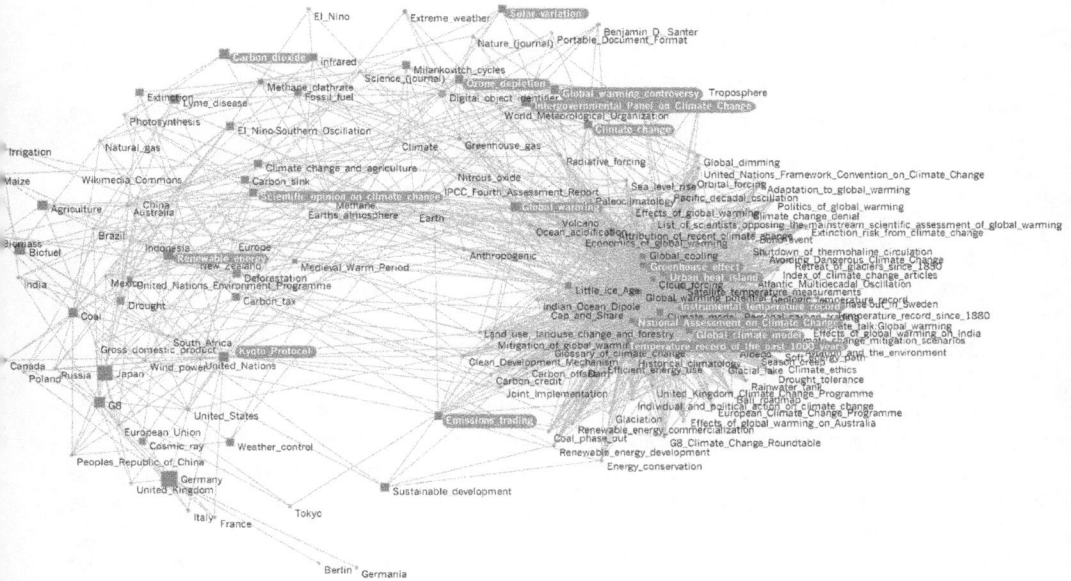

Figure 12: Article network graph. This graph depicts the network of Wikipedia articles inter-linked with the 'Global Warming' Wikipedia article. Nodes are sized according to numbers of links, shaped according to their role in the network (hubs appear in purple), and distributed according to their in- and out-degree centrality.[78]

One interpretation of the network of articles comes from the hypothesis that structurally, Wikipedia networks may represent the free encyclopedia's desire to resolve controversy (an aim embodied implicitly, for example, in the aforementioned NPoV core rule). From this per-spective, one sees a very clear separation — at the level of discourse and article delineations and links — of factual articles from articles dealing with the popular debate surrounding the existence and causes of Global Warming.[79] In further analyses (below), Stevenson and Gerlitz ask whether the creation of specific new articles dedicated to the controversy may be better viewed as a form of controversy management, one that is specific to Wikipedia.[80]

---

78   Digital Methods Initiative, 'The Place of Issues'.
79   In a brief study of the skeptics' resonance in this set of Wikipedia articles, I took the list of interlinked global warming-related articles and queried them for a list of known skeptics — the keynote speakers of the first Heartland climate change conference in 2008 — and found most mentions of these skeptics in the articles on the 'Climate Change Controversy' and the 'Inter-governmental Panel of Climate Change.' S. Fred Singer was the most mentioned skeptic, listed in four different global warming related articles. See also Digital Methods Initiative, 'Climate Change Sceptiks in the Wikipedia Climate Change Space'.
80   This relates to what Jemielnak phrased as dispute resolution mechanisms, and also to the sociological studies of science and technology as discussed in the first chapter.

Stevenson and Gerlitz commenced their study of 'controversy management' on Wikipedia by zooming in on editing activity within a select sample of articles address Global Warming. For each article in the sample, they tallied the number of edits per month from November 2001 to July 2009 and visualized this as a (over two meters wide) 'bubble line' heat map, where the intensity of the red color indicates editing activity (Figure 13).

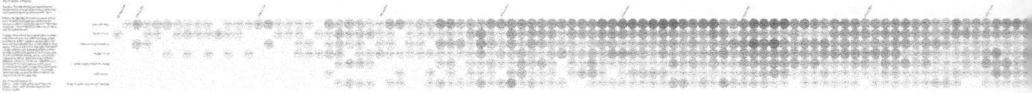

Figure 13: Editing heat map. This is an over two meters-wide bubble line heat map, visualizing the editing activity over time in a set of climate change-related articles. The intensity of the red color indicates the editing activity in the respective article.[81]

Networked Content Analysis allows for a historical reconstruction of a debate. Here, it appears to indicate generic Wikipedia editing trends, such as overall increases of editing interventions over time, and the relative decrease in activity in the months of June and December, as well as to mark out the existence of an 'incubation' period between an article's creation and its maturation, with initial editing and a period of inactivity followed by more regular editing. One may also recognize issue attention cycles as discussed in the Introduction, where 'new' news around the controversy or debate has the effect of spiking Wikipedia activity across specific pages. Accounting for tool-assisted human editors, who will receive alerts when 'their' articles have been edited, these upward spirals may have resulted in editing wars more than once. For example, consider the editing activity after the release of Climate Change 2007, the Fourth Assessment Report (AR4) of the United Nations Intergovernmental Panel on Climate Change (IPCC) in February of 2007. The sudden decline of activity for the 'Global Warming' and 'Global Warming Controversy' articles are the result of article protection of both articles after an editing war led administrators to close down the article from further editing. The heat map may thus also be used to signal significant moments in Wikipedia's management of the issue of global warming.[82]

In addition to this editing activity heat map, Gerlitz and Stevenson made a similar bubble heat map of bot activity. Here, they shift focus to the technical actors active in this article ecology and recognize two things. Firstly, and perhaps unsurprisingly, the most actively edited articles in the network have most bot activity. The four most active bots in this space (ClueBot, SmackBot, TawkerBot2, and AntiVandalBot) are anti-vandalism bots that are indeed also most active in the most-edited articles.[83][84] Secondly, the researchers found that bots do not

---

81   Digital Methods Initiative, 'The Place of Issues'..
82   Digital Methods Initiative, 'The Place of Issues'. Another part of their study zooms in on bot activity,
     which is similarly visualized as a heat map 'bubble line.'
83   ClueBot (now called ClueBot NG) is an anti-vandalism bot; SmackBot (presently called Helpful Pixie
     Bot) is an editing bot, mostly formatting articles. TawkerBot2 and its follow-up AntiVandalBot were anti-
     vandalism bots (currently inactive).
84   See also: Wikipedia contributors, 'User: AntiVandalBot', https://en.wikipedia.org/wiki/User:AntiVandalBot.

account for the high editing activity, as most bots that are editing these articles make only up to ten edits each.

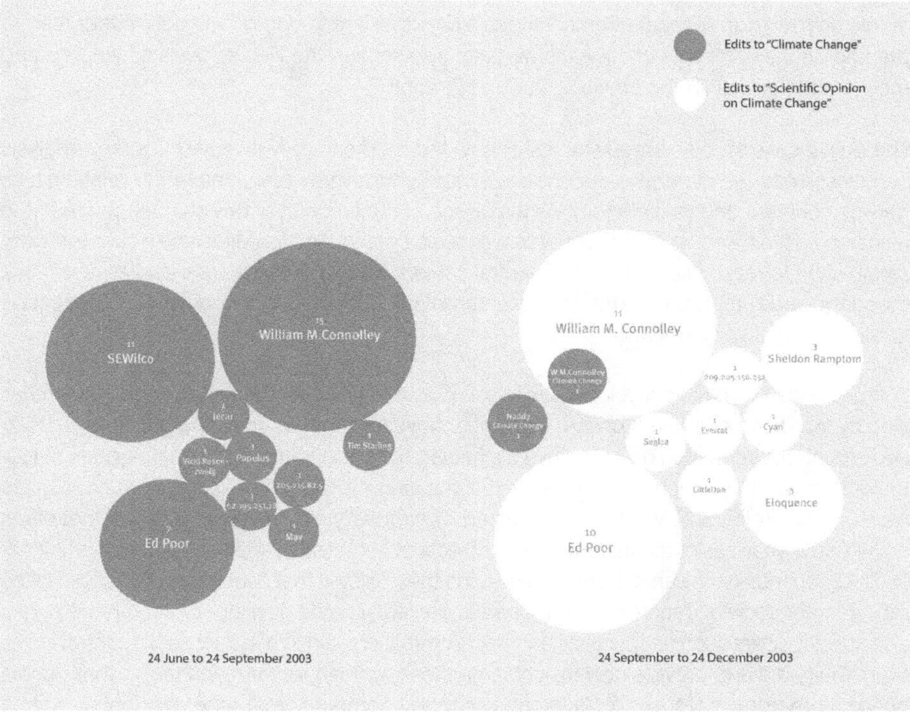

Figure 14: Editor migration map. This Dorling map visualizes the activity of editors active in the 'Climate Change' Wikipedia article in June of 2003 (left), as compared to those active in the articles on 'Climate Change' and 'Scientific Opinion on Climate Change' (right).[85]

More telling in this particular case is a closer view on actor editing activity in the context of controversy management on and by Wikipedia. In February of 2003, the article 'Scientific Opinion on Climate Change' was created, which has led to a decline in editing activity both in the article on 'Global Warming' and that on 'Climate Change'. By creating a separate article, the controversy was effectively displaced, taken out of the main articles, and as a 'controversy object' moved into its own dedicated space. Gerlitz and Stevenson looked close into this displacement by asking whether this displacement had let to editor migration from the main article on climate change to the controversy article on the scientific opinion on climate change. The visualization in Figure 14 shows the editing activity of those editors active in the 'Climate Change' article three months prior to the creation of 'Scientific Opinion' and three months after its creation. And indeed, we can see that most editors have migrated along with the newly created article, which again (just like the Gdańsk/Dantzig example) proves the effectiveness of this measure in the management of controversy on Wikipedia through forking. Only one of

85   Digital Methods Initiative, 'The Place of Issues'.

the editors active in the Climate Change article before the creation of the Scientific Opinion article remains active, however slightly, in the original article on Climate Change. The mass migration of editors who were active in the main article on climate change to the forked debate article on the issue yet again demonstrates a commitment to debate as such, rather than to the knowledge of climate change, as we have also seen in the web analysis of skeptics and their 'related issues' in the previous study in Chapter 3.

The above case studies are examples of how the methods of Networked Content Analysis can close-read the dynamics of controversy and controversy management in relationships between content and its technicity. As discussed in the first part of this chapter, much of the research has focused on the accuracy of Wikipedia content and its editor's collective (however small) effort to reach high quality and neutral content. However, these case studies reveal that for controversial topics, the articles presented may be the result of contestation, mediation, lock-down, or displacement.

Wikipedia, as an online encyclopedia project, presents hotly debated climate change entries side by side to more straightforward and uncontroversial entries. To further the study of Wikipedia content production and controversy, researchers, programmers, and designers of four universities working together in the context of the aforementioned project Electronic Maps to Assist Public Science (EMAPS) have created *Contropedia*, an analytical platform that offers novel visual analyses of the instances and objects of contestation within Wikipedia articles.[86] Their key orientation towards these inquiries and their utility is that conflicts on Wikipedia 'often reflect larger societal debates'.[87] Contropedia, presently being developed for both the public and specific users such as scientists and decision-makers, aims to extract social controversies from Wikipedia and provide new insights into these through visualization tools. Contropedia builds its metrics on those of Wikipedia itself, and combines real-time data about editing and discussion activity, to '[allow] for a deeper understanding of the substance, composition, actor alignment, trajectory and liveliness of controversies on Wikipedia'.[88] This commitment to the co-development of, essentially, a publicly available tool for networked content analysis, is perhaps a sign of this practice that I am outlining here starting to take form further, and is confirmed as necessary for public and civic sector needs. Contropedia is specific to Wikipedia, and could even help to refine the impact and relevance of the Wikipedia project, and will clearly provide a powerful tool for a networked content analysis of controversial issues, repurposing markers of technicity by reading them as markers of controversy (e.g. editing activity or talk page activity).

As discussed in the Introduction, in present media conditions, a clean separation of content from the platforms that serve and format it is no longer feasible. It is now impossible, or, at

---

86   Emaps, 'Contropedia'.
     My discussion of this project and other research and art projects related to big data was published in
     Big Data & Society: S. Niederer and R. Taudin Chabot, 'Deconstructing the Cloud: Responses to Big
     Data Phenomena From Social Sciences, Humanities and the Arts', *Big Data & Society* 2.2 (2015):
     http://doi.org/10.1177/2053951715594635.
87   Borra et al. 'Societal Controversies in Wikipedia Articles'.
88   Emaps, 'Contropedia'.

least, unadvisable, to regard a Wikipedia article as entirely separate from its publicly available production process. Questions regarding actor composition, bot activity, discussion, and forking are of great interest to those invested in content analysis in a networked era as such, and to anyone embarking on the mapping of a contemporary debate. Krippendorff has laid the groundwork for such analysis, well prior to content analysis having to deal with online content. Furthermore, Krippendorff has laid out the non-intrusiveness of the approach, the inclusion of content in all its shapes and forms, and the attention to the context of content, which are all applicable to the study of a debate in Wikipedia. By extending the approach to adapt to the specificities of networked content, I have proposed to take up digital methods and research *principles*, if you will, from controversy mapping. Herring, in her 2010 piece, has also suggested extending the paradigm of content analysis to suit web content. However, in contrast with her suggestion to pull in methods from non-digital realms, I propose to build on existing digital methods to suit the study of networked content. As controversy mapping urges researchers to *follow the actors* and *describe what you see* (rather than carrying pre-set categories and codebooks), this encourages the Networked Content Analysis researcher to make use of the networkedness of content and traverse content spaces.

Conclusions

In line with David Beer's call for a more thorough understanding of the 'technological unconscious' of participatory web cultures, I have in this chapter discussed several methods to study networked content while unraveling in detail the close interdependency of human and technological agents, in order to further the instruments needed for Networked Content Analysis.[89] It is important to comprehend the powerful information technologies that shape our everyday life and the coded mechanisms behind our informational practices and cultural experiences. The analysis of the Wikipedia platform as a socio-technical system is a first step in the direction of developing such adaptive techniques for networked content analysis.

The first generation of scholarly Wikipedia research has focused mainly on the platform's capacities for crowdsourcing knowledge production, as well as on the reliability of its co-produced content. I have argued for more attention to the machinery that facilitates and formats this knowledge production. While traditional content analysis may reach its limits to struggle with the omnipresence of technical agents in the wiki-platform of Wikipedia, networked content analysis provides means to properly assess Wikipedia's content, across articles and language versions. Nicolas Carr has compared Web 2.0 to the mechanical Turk (of the late 18th century), which 'turns people's actions and judgments into functions in a software program'.[90] Wikipedia, on the other hand, could be described as its opposite; people are so focused on watching the humans creating knowledge that they do not see the machinery and actual bots that are so entangled with what is created and collaborated.[91] A thorough and critical understanding of the automated processes that structure human judgments and decisions in

---

89   D. Beer, 'Power Through the Algorithm? Participatory Web Cultures and the Technological Unconscious', *New Media & Society*, 11.6 (2009): 985–1002.
90   Carr, *The Big Switch*, 218.
91   See also Niederer, 'Interview'.

and beyond online space requires analytical skills and medium-specific methods. These are crucial to a full understanding of how Wikipedia and other online platforms work. The methods are also useful for users learning to critically analyze their interactions with technology beyond softwarized modes of control, and towards active engagement in technologized knowledge development.[92] Furthermore, by assessing Wikipedia's content across articles and language versions, and its comparison to more static encyclopedia projects, frameworks, and tools for networked content analysis also make it clear how Wikipedia is socio-technically modulated towards reliability and consensus over time.

Wikipedia has never been an egalitarian space; its various user groups have very distinct levels of permissions, and it is not only human actors that form the hard core of editors. In this chapter, I have argued how Wikipedia's collaborative qualities and workings are complexly technical and hierarchical, involving not only human users but specific combinations of human and non-human actors.[93] Since 2002, Wikipedia content has been maintained by both tool-assisted human editors and bots, and collaboration has been modulated by protocols and strict managerial hierarchies. Bots are systematically deployed to detect and revert vandalism, monitor certain articles, and, if necessary, ban users, but they also play a substantial role in the creation and maintenance of content. As I have pointed out, bot activity may also be analyzed, perhaps counter-intuitively, as an indicator of the international or intercultural dimension of Wikipedia as a global project.

Studies that include technicity, non-human actors, and coded protocols can contribute greatly to our understanding of controversial topics such as climate change on platforms like Wikipedia. In this chapter, attention to climate change as a web-based controversy object, and to recent software projects such as Contropedia, enables a socio-technical view behind the scenes of collaborative knowledge production.[94] With its history tabs and discussion pages, its intricate administrative systems of editing policy, software robots, and tool-assisted humans, Wikipedia proves to be a place and platform par excellence to conduct networked content analysis to map controversy dynamics.[95]

Asking what kind of climate change debate Wikipedia puts forward, I want to conclude that Wikipedia offers critical insights into the socio-technics of online knowledge production and controversy management. However different its technicity is from other parts of the web, Wikipedia shares a capacity alongside the other platforms discussed here to be extremely useful for the study of actor commitment. The mass migration of editors of the main article

---

92   Zittrain, *The Future of the Internet*, 245.
93   Critiquing the presentation of non-human actors as existing more or less autonomously from human users, Jaron Lanier has argued that: 'Some people [...] believe they are hearing algorithms and crowds and other internet-supported nonhuman entities speak for themselves. I don't hear their voices, though – and I believe those who do are fooling themselves.' Lanier, *You Are Not a Gadget,* 39.
94   Climaps, 'Contropedia'.
95   R.S. Geiger and D. Ribes, 'The Work of Sustaining Order in Wikipedia: The Banning of a Vandal', in *Proceedings of the ACM 2010 conference on Computer supported cooperative work (CSCW)*, Atlanta, GA: Association for Computing Machinery, 2010, http://www.stuartgeiger.com/wordpress/wp-content/uploads/2009/10/cscw-sustaining-order-wikipedia.pdf.

on climate change to the forked debate article, for instance, yet again underlines the skeptics' commitment to debate *as such*, rather than to climate change as a specific topic and research field. This harkens back to the study of the skeptics on the web, where we found their 'related issues' to be largely unrelated to climate change (see Chapter 3). The different and recurring research findings, methodological insights, and analytics emphasized in this, and the previous chapter might prove to be scalable to other platforms and web infrastructures, too, as will be similarly explored in the following chapter on content networked by Twitter. In the next chapter, I will assess the composition of actors for even more specific climate-related discourses. Additionally, I will further 'profile' these sub-discourses by looking at most amplified content (retweets) and most-shared content (by looking at the URLs included in tweets). So far, I would argue that the vastly different technicities we have encountered in the first two case studies confirm the necessity to refine the definitions and demarcations of (the materiality of) content, and recognize the technicity as an active agent and part of networked content.

5. MAPPING THE RESONANCE OF CLIMATE CHANGE DISCOURSES IN TWITTER

In the previous chapters, I have proposed networked content analysis as an approach to the study of online networked content shaped by the technicity of its platforms and engines. The first case study traced climate change skeptics in science (through scientific publications) and on the web (looking at hyperlinking networks and their resonance in search engine results for the query of climate change). What I found was that networked content analysis presented the skeptics as *professional skeptics* engaged in skepticism of a variety of topics, rather than presenting them as scientists dedicated to the topic of climate change alone. The second case study discussed in detail the technicity of Wikipedia, as a socio-technical site especially suited for controversy mapping. The study of the climate debate in Wikipedia further estab-lished the profile of skeptics as dedicated to debate, as in the controversy management by Wikipedia editors, creating an article dedicated to the scientific debate, the actors most active in questioning and editing the article on climate change migrated along to the new article. They even never returned to the main article on the issue of climate change.

In this chapter, I will apply networked content analysis to the climate change debate in Twitter in the period of 2012-2014. More than in the previous case studies of the web and Wikipedia, I will discuss in detail the issue of climate change, its sub-issues, and the recent literature connecting it to conflict. This study entails working with the built-in logic of the platform and begins with recognizing the very particular (socio-technical) ways in which content is net-worked there. Therefore, I will first briefly discuss how content circulates on the micro-blogging platform.[1] This discussion is not designed to be a full glossary of Twitter features (which can be found on Twitter) but rather a brief introduction to the many ways in which content and its users are networked on Twitter.[2]

Twitter is a global messaging social network that allows its users to publish short messages (and links) up to 140 characters in length. These so-called 'tweets' can be posted by reg-istered users that have a username that starts with an @. Twitter prioritizes 'fresh' data and presents tweets in reverse chronological order (with the latest post on top) and does so in real-time.[3] For each tweet, Twitter displays some numeric data, such as the number of retweets and favorites, and a timestamp indicating how much time has passed since the tweet was posted. For each user, Twitter lists the number of followers, and the number of users this user is following, as well as the date of registration. Furthermore, users can add a short descrip-tion, a URL, and a location (even geo-location) to each of their tweets. All tweets are publicly accessible, except for direct messages between users and tweets from protected accounts.

---

1   For historical accounts of the development of the micro-blogging platform Twitter, see, for instance:
     Rogers, *Digital Methods*; Van Dijck, 'Tracing Twitter'; Van Dijck, 'The Culture of Connectivity'.
2   See, for instance, Twitter, 'Getting started with Twitter', https://support.twitter.com/articles/215585.
3   For a critical analysis of the freshness of data and the 'real-time-ness' of Twitter and other social media,
     see: Helmond, 'The Perceived Freshness Fetish' and Weltevrede, Helmond, and Gerlitz, 'The Politics of
     Real-time'.

Hashtags (keywords marked with a hash or #) are included in tweets to tag content and to participate in a public conversation, by connecting to public channels of content that carries the same hashtag (a convention to group content known from Internet Relay Chat [IRC]). Hashtags thus 'facilitate[s] a global discussion on a topic beyond a user's follower network', as they can be clicked to present a stream of all messages containing that hashtag (again, with the most recent tweet presented on top).[4] The use of hashtags can therefore also be interpreted as a willful means to connect to a broader conversation, trending beyond one's personal network. The use of hashtags for analysis has some limitations, as hashtags occur in less than 20% of all tweets and are used by specific users for specific practices.[5] However, as I will discuss later in this chapter, tweets containing multiple hashtags offer possibilities for co-hashtag analysis, where the co-occurring hashtags are regarded as topical clusters. Users following other users (to 'listen' to their stream of messages), is one of the prominent activities on Twitter.[6] This activity adds followed users' posts to one's own 'Timeline.' Other user interactions include @mentions (tweets that address a user by mentioning their @username), @replies (tweets sent in response to other tweets), and retweets.[7]

Retweeting, or the resending or quoting of another user's tweet, is done to amplify a message, sharing information with a user's followers, or commenting on a quoted message. Other motivations for retweeting are discussed extensively in boyd et al., based on interviews.[8] Retweeting has been built into the Twitter interface (alongside favorites and replies). Different third-party apps have different formats of retweeting, just as different users may style their retweets differently (for instance, by adding 'via @username' rather than RT, short for ReTweet), which should be taken into account when studying Twitter data.[9] Tweets may include URLs, where reported percentages of tweets with URLs vary from 22% to 11.7%.[10] Here, networked content analysts have to keep in mind that URLs may be shortened (for example, with bit.ly) in order to save space, i.e., a URL mentioned in a tweet can't always be recognized by a common web address including www.[11]

---

4   Lotan et al. 'The Revolutions Were Tweeted'.
      See also C. Gerlitz and B. Rieder, 'Mining One Percent of Twitter: Collections, Baselines, Sampling', *M/C Journal*, 16.2 (2013) for a discussion of the affordances of hashtags for research. They find, on the basis of their 1% sample analyzing 1 day of tweets, that 13,1% of the tweets include hashtags.

5   Gerlitz and Rieder, 'Mining One Percent of Twitter'.

6   Van Dijck, 'Tracing Twitter'.

7   danah boyd, S. Golder, and G. Lotan, 'Tweet, Tweet, Retweet: Conversational Aspects of Retweeting on Twitter', in *43rd Hawaii International Conference on System Sciences (HICSS)*, 2010, 2.
      See also Honeycutt and Herring, *Beyond Microblogging: Conversation and Collaboration via Twitter*, 2009, https://www.researchgate.net/publication/224373137_Beyond_Microblogging_Conversation_and_Collaboration_via_Twitter, for discussions of the various motivations users may have to include an @ mention, such as attention-seeking, addressing users, etc.

8   boyd et al. 'Tweet, Tweet, Retweet', 6.

9   Gerlitz and Rieder, 'Mining One Percent of Twitter', discuss demarcation of data in Twitter, as many case studies use specific hashtags or user practices (such as retweeting or favoriting) as a means to demarcate a sample, which is a question of recall (how many data points did I get?) and precision (how many of these data points are relevant?).

10   Smyrnaion and Rieder, 'Social Infomediation of News on Twitter'.

11   boyd et al. 'Tweet, Tweet, Retweet', 2.

As tweets can cover all sorts of mundane topics, but also carry more substantive missives of public political and informational value, the use of Twitter data for scholarly research is becoming widespread.[12] According to Tumasjan et al., tweets can function as indicators of political opinion, while Twitter offers a platform for political deliberation, which also makes it a highly suitable site for controversy analysis around a social issue.[13] [14] The choice of including Twitter as a platform for the study of the form and substance of the issue of climate change and vulnerability concepts therein is not arbitrary. Twitter relates knowledge perception, reception, and conversation. Furthermore, Twitter has an interesting relationship with mass media content, as it is not just a media platform, but a platform that transpires within multiple media networks. Twitter could be approached through more conventional news cycle analyses but also through 'meme-tracking'.[15] In the latter mode, Twitter as a micro-blog could then be seen as highly responsive to or even parasitical or imploding of conventional news 'sites', echoing and amplifying news snippets by tweeting and retweeting. Further, as Twitter is often moving information faster than the news, Twitter content, in some cases, *is* news. Of course, for these reasons, Twitter is a popular medium for professional journalists. They bind tweets to their stories, and when their work has been published, they may tweet a link to that article, using it as a channel for the distribution of their own work. As news and mass media sources strive to make their content 'platform-ready', a term by Helmond, the entanglement of news, other mass media content, and new platforms has entered the next level.[16] Networked content analysis proposes to take this entanglement as a given and to demarcate content through the logic of the platform (as developed in digital methods) and thus follow the actors across sources (as is key to controversy analysis). The rise of digital media does not mean the end of traditional mass media, but its reconfiguration as part of online networked content. This is

---

12   The use of Twitter data for cultural and social analysis has been described as the third phase in Twitter's popular cultural uptake, which had as its first phase the function of being an 'ambient friend-following tool', where user content answers the question 'What are you doing?' The second phase of Twitter usage encouraged by Twitter's new tagline 'What's happening?' both recognized and further fostered its use as a 'news medium for event-following'. Rogers, R. 'Debanalising Twitter', xii-xiv.

13   A. Tumasjan, T.O. Sprenger, P.G. Sandner, and I.M. Welpe, 'Predicting Elections with Twitter: What 140 Characters Reveal About Political Sentiment', in *Fourth International AAAI Conference on Weblogs and Social Media*, 2010, https://www.aaai.org/ocs/index.php/ICWSM/ICWSM10/paper/view/1441. In their 2010 study, Tumasjan et al. studied deliberation by looking at the exchange of substantive issues and equality of participation (as put forward by Koop and Jansen in their study of blogs as sites of deliberation). Through a content analysis of 100,000 tweets about German political parties around the federal elections of 2009, they found that Twitter was used extensively for political deliberation, with a massive number of tweets mentioning one or more of the political parties, and one-third of these messages partaking in platform-based conversations.

14   The predictive affordances of Twitter have been criticized by scholars such as Daniel Gayo-Avello, whose paper from 2012 offers an interesting 'annotated biography' with a discussion of Twitter prediction literature. D. Gayo-Avello, 'I Wanted to Predict Elections with Twitter and All I Got Was This Lousy Paper: A Balanced Survey on Election Prediction Using Twitter Data', *Arxiv Preprint arXiv12046441*, 2012, http://arxiv.org/pdf/1204.6441.pdf.

15   J. Leskovec, L. Backstrom, and J. Kleinberg, 'Meme-tracking and the Dynamics of the News Cycle', In *Proceedings of the 15th ACM SIGKDD international conference on Knowledge discovery and data mining*, ACM, 2009, pp. 497-506, http://dl.acm.org/citation.cfm?id=1557077.

16   Helmond, *The Web as Platform*.

important to bear in mind analytically, and a key to its utility for research practices such as networked content analysis.

Just as in Wikipedia and the web (accessed through Google Web Search), it is no longer possible to separate content from its carrier. Looking at the entanglement of content with Twitter's technicities of distributing, networking, and amplification its content, it also highly unadvisable to even attempt to ignore these mechanisms.[17] Taking that as a starting point of networked content analysis, where any evaluation of online content should acknowledge the significance of its socio-technological structure, I operationalize the previously introduced socio-technics of Twitter (in shared links, retweets, etc.) in the following analyses of the climate change debate. Firstly, I will compare the resonance of terms associated with climate change, including skepticism, mitigation, adaptation, and conflict through a climate change content collection in Twitter. This is to propose that the changing prominence of each concept in time indicates a 'phase' in the issue evolution of climate change as a controversy object.[18]

For the first part of the case study, I worked with a data set containing 8.3 million climate change tweets (from the period of 23 November 2012 until 30 May 2013), which I queried for the keywords [skeptic], [mitigation], [adaptation] and [conflict OR violence], using the online Digital Method Initiative's Twitter Capture and Analysis Tool (DMI-TCAT).[19] Each of these queries refer to one of four related climate change discourses: skepticism (towards the man-made origins and unprecedentedness of climate change), mitigation (the prevention of further climate change by minimizing its causes), adaptation (to climate change), and conflict (here taken to mean political unrest relatable to climate change vulnerability).[20] Given 'vulnerability' has become a prominent and focalizing, contested discourse within climate change debates, both in the scientific literature (as mapped out by the IPCC in 2014) and in news coverage around climate change, I will discuss this more elaborately.[21] Here, I will build on the influential work of sociologist Ulrich Beck, who has described climate change as one of

---

17 DMI-TCAT, as a tool, does separate Twitter content from the platform Twitter. However, it retains information about how Twitter structures its information.
18 EMAPS, 'Vulnerability, Resilience and Conflict'. Needless to say, these phases are not cleanly separated chronologically but rather overlap.
19 Borra and Rieder, 'Programmed Method'.
See also E. Borra and B. Rieder, 'Programmed Method: Developing a Toolset for Capturing and Analyzing Tweets', *Aslib Journal of Information Management*, 66.3 (2014): 262–278.
20 In the EMAPS Digital Methods Fall Data Sprint, we also asked whether conflict could be seen as a fourth phase in the evolution of the issue of climate change, after skepticism, mitigation, and adaptation. EMAPS, 'Vulnerability, Resilience and Conflict'.
21 The IPCC's Working Group II has mapped adaptation within scientific literature on climate change and concludes that there is an overall doubling of the volume of publication in this field in less than five years, and secondly, that adaptation has become a central area of research within the scientific literature on climate change. EMAPS, 'Reading the State of Climate Change From Digital Media', 2014, http://climaps.eu/#!/narrative/reading-the-state-of-climate-change-from-digital-media.
See also IPCC, 'Climate Change 2014: Synthesis Report: Contribution of Working Groups I, II and III to the Fifth Assessment Report of the Intergovernmental Panel on Climate Change', Geneva: IPCC, 2014, http://ar5-syr.ipcc.ch/, 3.

the main problems of our *World at Risk*.[22] In his framing, multiple anticipated crises (climate change, terrorism, financial disaster, and so on) lead to a situation in which:

> The decoupling of the social location and the social decision-making responsibility from the places and times in which other "foreign" populations become (or are made) the object of possible physical and social injuries.[23]

This decoupling between the decision-making and the sites of such possible 'injuries', or *casualties*, can be clearly demonstrated when looking at the assessment of climate change adaptation and climate change vulnerability, and the way discussions about the distribution of resources to those places most vulnerable to the adverse effects of climate change play out at the UN Framework Convention on Climate Change's Conference of the Parties (UNF-CC COP).[24] Climate change vulnerability, according to the IPCC, is the 'degree to which a system is susceptible to and is unable to cope with adverse effects (of climate change)'.[25] Vulnerability research is, therefore, interested in 'the shocks and stresses experienced by the social-ecological system, the response of the system, and the capacity for adaptive action'.[26]

The Kyoto protocol's Adaptation Fund and the UNFCC have described their commitment to and funding of adaptation as designed 'to assist developing countries that are particularly vulnerable to the adverse effects of climate change'.[27] Importantly, the assessment of such *particularly vulnerable* countries has been critically described as a 'political challenge', rather than a scientific effort, as the socio-economic variables addressed when determining vulnerability blur the line between adaptation actions and development aid.[28] The prominence that is now given to 'adaptation' and 'vulnerability' discourses and models within the discussion of climate change, both in the UNFCC and scientific literature and on an operational level, as in the field of urban planning, has led to the declaration of an 'adaptation turn'.[29]

The following case study addresses the further development of Networked Content Analysis by attending to technicities of the widely used and globally accessed Twitter platform. The

22 Beck, *World at Risk*.
23 Beck, *World at Risk*, 161.
24 Beck, *World at Risk*, 161.
25 IPCC, 'Working Group II: Impacts, Adaptation and Vulnerability: Summary for Policymakers', 2001, http://www.ipcc.ch/ipccreports/tar/wg2/index.php?idp=8.
26 Adger, 'Vulnerability', 269.
27 R.J.T. Klein, 'Identifying Countries That Are Particularly Vulnerable to the Adverse Effects of Climate Change: An Academic or a Political Challenge?' *Carbon and Climate Law Review* 3 (2009): 284–289.
28 Klein, 'Identifying Countries That Are Particularly Vulnerable to the Adverse Effects of Climate Change', 289. This is discussed in detail in EMAPS, 'Who Deserved to Be Funded? A Closer Look at the Practices of Vulnerability Assessment and the Priorities of Adaptation Funding', 2014, http://climaps.eu/#!/narrative/who-deserves-to-be-funded.
29 J. Howard, 'Climate Change Mitigation and Adaptation in Developed Nations: A Critical Perspective on the Adaptation Turn in Urban Climate Planning', in S. Davoudi, J. Crawford and A. Mehmood (eds) *Planning for Climate Change: Strategies for Mitigation and Adaptation for Spatial Planning*, London: Earthscan, 2009, pp. 19-32.
See also Venturini et al. 'Climaps by EMAPS in 2 Pages'.

case study foregrounds not just the utility of Twitter for such analyses but also, in the other direction, considers which of the aforementioned networked content analysis methods and techniques developed in the previous case studies (in chapters 3 and 4) might also be applied to the platform of Twitter, and which others are so productively platform-specific to be non-transferable.

Using Twitter Data for Research

Twitter has often been described as an important channel during political events and social unrest.[30][31][32] At the same time, popular and scholarly assessments of the role played by Twitter in social uprisings come with some caveats. For example, news coverage of the uprisings in Iran has been (productively) criticized as 'heavily skewed' towards being presented as a technology-driven social movement.[33] Gladwell has pointed out that such skewing is due partly to Western scholars' and media pundits' own 'outsized enthusiasm(s) for social media'.[34] Other scholars have looked closer at the composition of the actors in the various uprisings, painting a more fine-grained picture of the role and relevance of the platform in these uprisings.[35]

According to Hermida, Twitter is a site for 'the immediate dissemination of digital fragments of news and information from official and unofficial sources over a variety of systems and devices', and might, therefore, be better understood as an 'awareness system', rather than merely a micro-blogging platform.[36] This awareness system functions as an always-on communication channel, ready to move 'from the background to the foreground' when necessary.[37] Twitter, Hermida argues, creates the means for 'ambient journalism', where value does not lie in any single tweet but rather in the 'awareness system that offers diverse means to collect, communicate, share and display news and information, serving diverse purposes'.[38] And it is this function of Twitter as an awareness system that I will assess in the case study of Twitter hashtag clusters.

Twitter has been analyzed as a source of *happening content* and *fresh data*, as a site for *real-time research*, as a platform with a 'dual nature of information source and conversation

---

30   Shirky, *Here Comes Everybody.*
31   Sullivan, 'The Revolution Will Be Twittered'.
32   Z. Tufekci and C. Wilson, 'Social Media and the Decision to Participate in Political Protest: Observations from Tahrir Square', *Journal of Communication* 62.2 (2012): 363–379.
33   E. Morozov, 'Iran: Downside to the "Twitter Revolution"', *Dissent* 56.4 (2009): 10–14.
34   Gladwell, 'Small Change'.
35   T. Poell and K. Darmoni, 'Twitter as a Multilingual Space: The Articulation of the Tunisian Revolution Through #sidibouzid', *NECSUS European Journal of Media Studies* 1.1 (2012): 14–34.
36   The term 'awareness systems' here refers to systems that support remote co-working. A. Hermida, 'Twittering the News', *Journalism Practice* 4.3 (2010): 298-301.
37   Hermida, 'Twittering the News', 298.
38   Hermida, 'Twittering the News', 301.

enabler'[39], and as an (archived) data set as well as and anticipatory medium.[40][41][42] Methods and tools for capturing and analyzing this real-time data have been developed for instance by Bruns and Liang, who study Twitter as an important channel for crisis communication during and after natural disasters, and by scholars who have looked at the predictive quality of tweets in relation to the stock market, such as Sprenger et al., or political sentiment around elections, such as Tumasjan et al.[43][44][45]

In what follows, I will look at the content that Twitter serves around the issue of climate change, and conduct a Networked Content Analysis of a year's worth of English-language climate-related tweets, exploring the 'Twitter ecology' of climate change content.[46] Twitter evidently does not produce 'climate science' but instead, puts scientific research into circulation while enabling up close, located and platform-literate engagements able to assess the resonance of climate change adaptation and indicators of vulnerability within the broader online discussion of climate change. Before exploring the resonance of the adaptation turn on Twitter, I will discuss the critical need to attend to vulnerability and adaptation concepts through a review of recent literature (news media, NGO reports, and scientific literature) that is connecting the risk of climate change to injuries and to conflict.[47] Combining a description of vulnerability assessments from published reports and media content with a methodological application of digital methods to Twitter, this chapter shows networked content analysis working to unpack and give analytic complexity to important discourses *within* the issue of climate change. This chapter focuses on the period of 2012-2014, a timeframe during which conflict was increasingly attributed to climate change, as I will discuss in the next section.

Climate Change Vulnerability and Its Relation to Conflict

Climate scholar Richard Klein has recently paid due critical attention to this rise of vulnerability research in scientific work. Klein describes how 'vulnerability has become a popular concept in a very diverse set of research fields' in projects ranging from 'studies of vulnerability to terrorism, to poverty, to computer viruses, to oil spills, to globalisation, to radiation, to SARS, to earthquakes, to financial collapse, to political change, and so on'.[48] Importantly,

---

39   G. Veltri, 'Microblogging and Nanotweets: Nanotechnology on Twitter', *Public Understanding of Science* 22.7 (2013): 832–849.

40   Rogers, 'Debanalising Twitter', xiv.

41   Back et al. 'Doing Real Time Research'.

42   Marres and Weltevrede, 'Scraping the Social?'

43   T.O. Sprenger, A. Tumasjan, P.G. Sandner, and I.M. Welpe, 'Tweets and Trades: The Information Content of Stock Microblogs', *European Financial Management* 20 (2014): 926-957, 10.1111/j.1468-036X.2013.12007.x.

44   Tumasjan, Sprenger, Sandner, and Welpe, 'Predicting Elections with Twitter'.

45   A. Bruns and Y.E. Liang, 'Tools and Methods for Capturing Twitter Data During Natural Disasters', *First Monday* 17.4 (2012): http://firstmonday.org/ojs/index.php/fm/article/view/3937/3193.

46   boyd et al. 'Tweet, Tweet, Retweet'.

47   In this chapter, more than in the previous case studies on the web and Wikipedia, I will discuss in detail the issue of climate change, its sub-issues, and the recent literature connecting it to conflict.

48   Klein, 'Identifying Countries That Are Particularly Vulnerable to the Adverse Effects of Climate Change', 285.

the particular connection I want to make between climate vulnerability and conflict has been steadily gaining attention in both scholarly research and popular media outlets. Following the publication of a research article on climate and conflict by Hsiang, Burke and Miguel, media outlets themselves began to pose speculative research questions,[49] for example: 'Could hotter temperatures from climate change boost violence?' and, 'How could a drought spark a civil war?'[50] The link between the Arab Spring and climate change was quickly made during this time, as headlines reported 'Drought helped cause Syria's war. Will climate change bring more like it?' and 'Climate change and rising food prices heightened Arab Spring'.[51]

The climate-conflict nexus, however, comprises many complicated facets of indexing and data triangulation, spurring further debates among scientists within and across disciplines.[52] [53][54] The emerging literature on climate change and conflict further appears to focus on two broader questions: 'how' climate change leads to conflict and 'where' climate change-induced conflicts will most likely take place.[55] As bleak headlines already indicate, a variety of climatic

---

49   S.M. Hsiang, M. Burke, E. Miguel, 'Quantifying the Influence of Climate on Human Conflict', *Science*, 341.6151 (2013): http://doi.org/10.1126/science.1235367.

50   Doucleff, 'Could Hotter Temperatures from Climate Change Boost Violence?'. NPR, 'How Could a Drought Spark a Civil War?'

51   Perez, 'Climate Change and Rising Food Prices Heightened Arab Spring'. Plumer, 'Drought Helped Cause Syria's War'.

52   A.T. Bohlken and E.J. Sergenti, 'Economic Growth and Ethnic Violence: An Empirical Investigation of Hindu-Muslim Riots in India', *Journal of Peace Research*, 47 (2010): 589–600.

53   C.S. Hendrix and I. Salehyan, 'Climate Change, Rainfall, and Social Conflict in Africa', *Journal of Peace Research* 49.1 (2012): 35–50.

54   Other recent literature indicates that in low-income settings, extreme rainfall events that adversely affect agricultural income are similarly associated with higher rates of personal violence and property crime. D. Blakeslee and R. Fishman, 'Rainfall shocks and property crimes in agrarian societies: Evidence from India', 2013, http://papers.ssrn.com/sol3/papers.cfm?abstract_id=2208292. H. Mehlum, E. Miguel, and R. Torvik, 'Poverty and Crime in 19th Century Germany', *Journal of Urban Economics* 59.3 (2006): 370–388.
     Some longitudinal studies of intergroup violence point out that such social conflicts tend to be more likely after extreme rainfall conditions. Reduced agricultural production may be an important mediating mechanism of conflict, although alternative explanations such as political instability cannot be excluded. On a local level, several studies in psychology and economics have found that individuals are more likely to act aggressively or show violent behavior if ambient temperatures at the time of observation are higher. C.A. Anderson, 'Heat and violence', *Current Directions in Psychological Science*, 10.1 (2001): 33–38. A. Auliciems and L. DiBartolo, 'Domestic Violence in a Subtropical Environment: Police Calls and Weather in Brisbane', *International Journal of Biometeorology*, 39.1 (1995): 34–39. D.T. Kenrick and S.W. MacFarlane, 'Ambient Temperature and Horn Honking: A Field Study of the Heat/Aggression Relationship', *Environment and Behavior* 18 (1986): 179–197.
     It is important to clarify how 'solid' this relationship between climate change and conflict is conceived to be at the time of writing. In a meta-analysis conducted by Hsiang, Burke, and Miguel 'Quantifying the Influence of Climate on Human Conflict', who evaluated 60 primary studies on the topic, particular trends are observed. For one, deviations from average rainfall and temperatures, whether up or down, are likely to result in human conflict on three levels, from the more local level of interpersonal violence and crime, moving to intergroup violence and political instability, and then measuring conflict at the global level, in terms of institutional breakdown and the collapse of civilizations. 'Quantifying the Influence of Climate on Human Conflict', 1.

55   T.F. Homer-Dixon, 'On the Threshold: Environmental Changes as Causes of Acute Conflict', *International Security* 16 (1991): 76–116.

variables are considered to be of influence on human conflict. According to Barnett and Adger, there are two ways in which conflict might be stimulated by climate change.[56] First, in line with research by Rifkin, changes in the political economy of energy resources (due to mitigative action to reduce emissions from fossil fuels) could result in conflict.[57] Second, conflict could be stimulated by the effects of actual or perceived long-term or short-term climate impacts in causing changes to social systems.[58] Short-term impacts include a change in the intensity and frequency of floods, droughts, storms and cyclones, fires, heatwaves, and epidemics. In the long term, changes in average conditions such as temperature, sea level, and annual precipitation will impact social-ecological systems. Also mediating the relationship between these climatic changes and human conflict are the interrelated issues of resource scarcity (cropland, freshwater, fisheries or forests) and migration.[59] Environmentally induced migration could lead to increased pressures on resources in areas or countries of destination and inter-communal tensions in source areas.[60][61] These trends may also complicate future food security as the competition around increasingly scarce resources proliferates.

The question then of where climate change-induced conflicts (and other casualties and damages) will most likely take place, makes the question of how the concept of climate vulnerability can be studied even more urgent. A number of studies on the connection between climate change and conflict note that the vulnerability of people to climate change depends on the extent to which they are dependent on natural resources and ecosystem services, the extent to which the resources and services they rely on are sensitive to climate change, and their capacity to adapt to changes in these resources and services.[62] Furthermore, those countries that do not have the ability to adapt to environmental change — often poor and underdeveloped states — are, in turn, more vulnerable to environmentally-related violence.[63] This vulnerability to climate change impacts and related effects such as violence is described in terms of a lack of 'adaptive capacity', or 'the ability or potential of a system to respond successfully to climate variability and change. [...] Common traits include human and social capital, wealth, technology, and the quantity and quality of infrastructure.'[64] These traits are among the variables used in so-called climate vulnerability indices, published as annual research reports that rank countries according to their adaptive capacity to climate change.

56   J. Barnett and W.N. Adger, 'Climate Change, Human Security and Violent Conflict', *Political Geography*, 26.6 (2008): 639–655.
57   Jeremy Rifkin, *The Hydrogen Economy: The Creation of the Worldwide Energy Web and the Redistribution of Power on Earth Tarcher* (New York: Putnam, 2002).
58   Barnett and Adger, 'Climate Change, Human Security and Violent Conflict'.
59   C. Raleigh and H. Urdal, 'Climate Change, Environmental Degradation and Armed Conflict', *Political Geography* 26.6 (2007): 674–694.
60   J. Barnett, 'Security and Climate Change', *Global Environmental Change*, 13.1 (2003): 7–17.
61   R. Reuveny, 'Climate Change-induced Migration and Violent Conflict', *Political Geography* 26.6 (2007): 656–673.
62   Adger, 'Climate Change, Human Security and Violent Conflict'.
63   Homer-Dixon, 'On the Threshold'.
64   E.A. Stanton, J. Cegan, R. Bueno, and F. Ackerman, *Estimating Regions' Relative Vulnerability to Climate Damages in the CRED Model*, Somerville, MA: Stockholm Environment Institute, 2011, 4.

Vulnerability Indices and the Assessment of Adaptive Capacity

Since the 1990s, there have been many projects that attempted to develop indices that claim to measure vulnerability to social and environmental change.[65] These vulnerability indices typically combine multiple indicators of a variable into a single measure, thus ordering a set of entities into quantitative attributes or traits. As such, they are integral to many contexts that require systematic approaches to decision-making, especially those that concern the management or governance of risk.[66] [67] At the same time, however, according to Barnett, Lambert and Fry, there have been so many attempts to create such indices that it has '[lead] the National Research Council (2000) to conclude that there is no consensus on their appropriateness, theoretical and scientific basis, and appropriate level of specificity or aggregation'.[68] Furthermore, measuring vulnerability has been described as 'impossible', as well as problematic in 'rais[ing] false expectations', around socio-ecological systems, given that 'there is ambiguity on what exactly the problem to be solved is and no canonical solution exists'.[69] [70] Nevertheless, vulnerability research aims to inform decision-making around funding opportunities to mitigate the worst possible impacts of climate change for particularly vulnerable target nations.[71] [72]

---

65 J. Barnett, S. Lambert, and I. Fry, 'The Hazards of Indicators: Insights from the Environmental Vulnerability Index', *Annals of the Association of American Geographers*, 98.1 (2008): 102–119.

66 Beck, *World at Risk*.

67 O. Renn and P. Graham, *White Paper on Risk Governance: Towards an Integrative Approach*, International risk governance council, 2015.

68 Barnett, Lambert, and Fry, 'The Hazards of Indicators: Insights from the Environmental Vulnerability Index', 106.

69 J. Hinkel presents an analysis of six diverse types of problems that vulnerability indicators are meant to address according to his review of the literature: '(i) identification of mitigation targets; (ii) identification of vulnerable people, communities, regions, etc.; (iii) raising awareness; (iv) allocation of adaptation funds; (v) monitoring of adaptation policy; and (vi) conducting scientific research'. Based on this, he finds that only the second type of problem can be addressed by vulnerability indicators, but only at small and local scales, causing him to question the concept of vulnerability itself and the applied methodologies. J. Hinkel, 'Indicators of Vulnerability and Adaptive Capacity: Towards a Clarification of the Science–policy Interface', *Global Environmental Change* 21.1 (2011): 198-206.

70 S.H. Eriksen and P.M. Kelly, 'Developing Credible Vulnerability Indicators for Climate Adaptation Policy Assessment', *Mitigation and Adaptation Strategies for Global Change* 12.4 (2007): 495–524.

71 Naomi Klein, in her book *This Changes Everything* (2014), discusses this as a justice issue. Many developing countries, due to both their specific local environments and limited infrastructures, will be worse hit by the impacts of climate change while having contributed least (e.g., in the sense of carbon emission levels) to creating the problem in the first place.

72 In his review of vulnerability research traditions, climate change scholar W. Neil Adger distinguishes between two scholarly 'antecedents' that have 'acted as seedbeds for ideas that eventually translated into current research on the vulnerability of social and physical systems in an integrated manner'. These are 'the analysis of vulnerability as lack of entitlements and the analysis of vulnerability to natural hazard'. This double-ness in the history of the research concept has lead to distinct parallelism in research practices where some researchers focus solely on ecological systems and 'largely ignore physical and biological systems (entitlements and livelihoods, while others 'try to integrate social and ecological systems'. A serious challenge following from the rise of adaptation and its inherent complexity is the question of how to develop robust and credible indicators and criteria for measuring vulnerability. Adger, 'Vulnerability', 270.

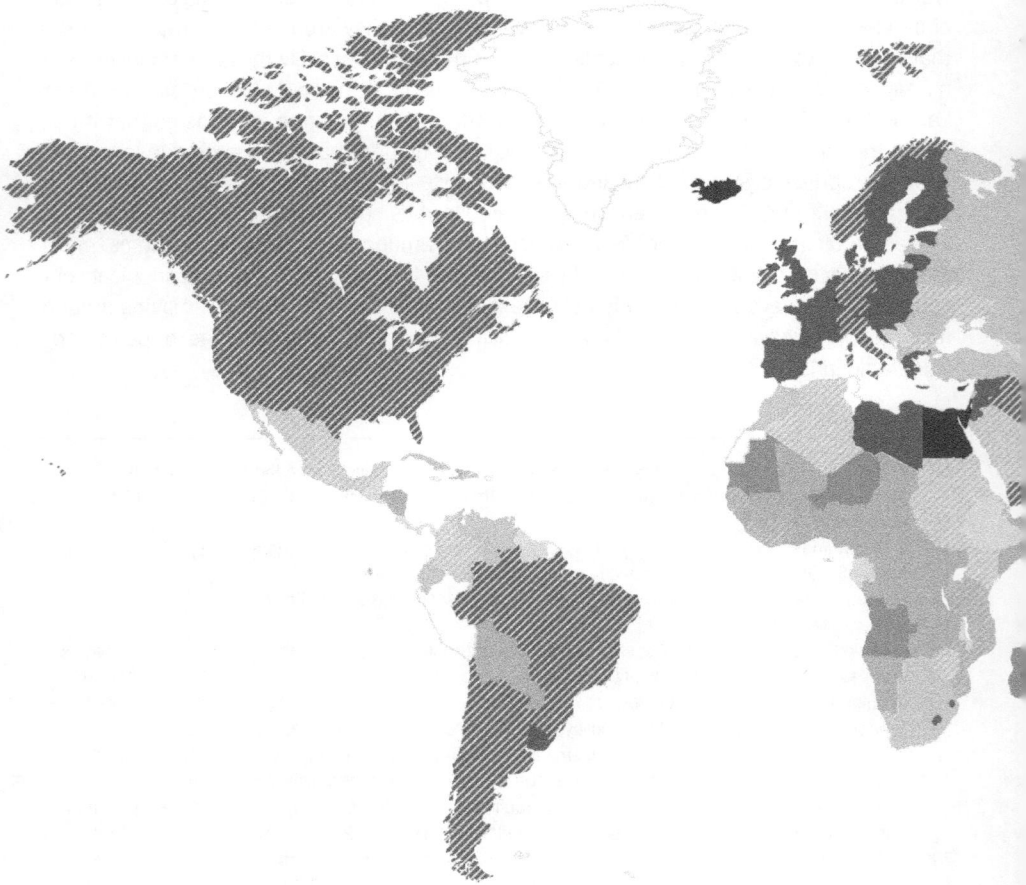

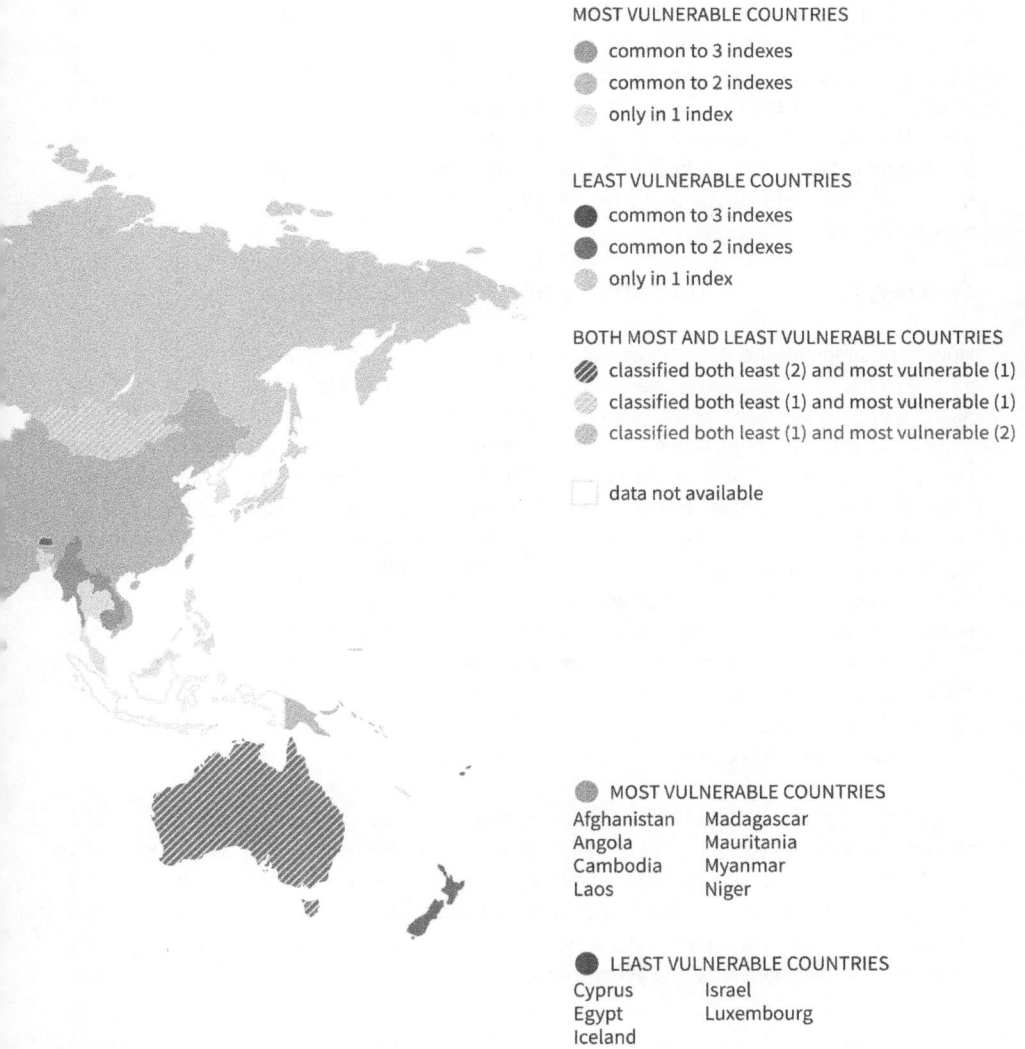

MOST VULNERABLE COUNTRIES

⬤ common to 3 indexes
⬤ common to 2 indexes
⬤ only in 1 index

LEAST VULNERABLE COUNTRIES

⬤ common to 3 indexes
⬤ common to 2 indexes
⬤ only in 1 index

BOTH MOST AND LEAST VULNERABLE COUNTRIES

⬤ classified both least (2) and most vulnerable (1)
⬤ classified both least (1) and most vulnerable (1)
⬤ classified both least (1) and most vulnerable (2)

▢ data not available

⬤ MOST VULNERABLE COUNTRIES

Afghanistan	Madagascar
Angola	Mauritania
Cambodia	Myanmar
Laos	Niger

⬤ LEAST VULNERABLE COUNTRIES

Cyprus	Israel
Egypt	Luxembourg
Iceland	

Figure 15: Who is vulnerable according to whom? This world map visualizes an exploratory comparative analysis of Germanwatch's Climate Risk Index (CRI), DARA's Climate Vulnerability Monitor (CVM), and the Global Adaptation Initiative's Global Adaptation Index (GAIN) in their assessment of vulnerability.[73]

---

73 EMAPS, 'Who is Vulnerable According to Whom?'.
    See also: http://climaps.org/?utm_content=buffer51f08&utm_medium=social&utm_source=twitter.
    com&utm_campaign=buffer#!/map/who-is-vulnerable-according-to-whom.

In a comparative analysis of three vulnerability indices, their ranked lists of most and least vulnerable countries and their usage, we have found that countries calculated to be most vulnerable and at-risk according to one Index may be among those with the greatest adaptive capacity according to the other Indices that take into account other variables.[74] Figure 15 (p. 100-101) shows a world map that compares the output of this triangulation, which illustrates some comparative appreciation of vulnerability, but also the lack of consensus on methodologies and, therefore, rankings of vulnerability. It is not surprising that the assessment of climate change vulnerability using indicators continues to divide both policy and academic communities alike.[75]

Twitter, Climate Vulnerability and the Adaptation Turn

However significant the differences between the three discussed indices may be, the lack of consensus does not seem to have hindered the coverage and talk of adaptation in official negotiations and gatherings as well as scientific literature, where a turn of attention to climate adaptation has been recognized. In the remainder of this chapter dedicated to Twitter, I will ask what kind of view on the climate change debate Twitter enables. Does a climate change *awareness system* indeed play out through the platform? And secondly, does an adaptation turn resonate here too? Taking as a starting point of Networked Content Analysis, the notion that any evaluation of online content should acknowledge the significance of its socio-technological structure, I operationalize the previously introduced socio-technics of Twitter — in shared links, retweets, etc. — in the following analyses of this case study. Firstly, I will compare the resonance of terms associated with climate change, including skepticism, mitigation, adaptation, and conflict through a climate change content collection in Twitter. This is to propose that the changing prominence of each concept in time indicates a 'phase' in the issue evolution of climate change as a controversy object.[76]

As mentioned earlier in this chapter, for this analysis a data set containing 8.3 million climate change tweets (from the period of 23 November 2012 until 30 May 2013) is queried for the keywords 'skeptic,' 'mitigation,' 'adaptation' and 'conflict OR violence,' using the online Twitter Capture and Analysis Tool (TCAT).[77] Following the logic of the Twitter platform, I have created profiles for each keyword indicating their various socio-technical formats of resonance, listing their URLs, top 10 hashtags, top 10 mentioned users, top 10 active users, and top 10 hosts

---

74   EMAPS, 'Who is Vulnerable According to Whom?', 2014, http://climaps.eu/#!/map/who-is-vulnerable-according-to-whom.
75   Hinkel, 'Indicators of Vulnerability and Adaptive Capacity', 198.
76   EMAPS, 'Vulnerability, Resilience and Conflict: Mapping Climate Change, Reading Cli-fi', *Electronic Maps to Assist Public Science Blog*, 2013, http://www.emapsproject.com/blog/archives/2293.
77   Borra and Rieder, 'Programmed Method'.
      The climate change collection was made with TCAT by collecting tweets that mention climate change (also spelled as climatechange), global warming (and globalwarming), climate, drought, or flood. This is a very wide data set, opting for high recall and low precision, which we then filtered, retaining only tweets mentioning 'climate change' or 'global warming.' The data set is available from the tool at: http://tcat.digitalmethods.net/analysis/index.php?dataset=globalwarming&query=&url_query=&exclude=&from_user_name=&from_source=&startdate=2012-11-23&enddate=2013-05-30&whattodo=&graph_resolution=day.

(of the URLs mentioned in the tweets). The profiles include the most linked URL and the most retweeted tweet for each of the keywords. Focusing on the top does not merely attune to the logic of the platform and its ranking; from a user perspective, it means the selection of content with the most exposure, those tweets most viewed by users.[78]

Figure 16 (p. 104) offers a visual rendition of these discourse profiles. The term of 'adaptation' resonates most in the climate change tweets, with 30,560 results, indicating that indeed also in Twitter the 'adaptation turn' has occurred.[79] Overall, 'adaptation' and 'mitigation' have similarities in terms of most-used hashtags (where five hashtags from the top 10 are shared). The occurrence of 'adaptation' in the 'mitigation' set and vice versa further confirms the overlap between the two terms. The UN and its events dominate both 'mitigation' and 'adaptation' (e.g., the users UN climate talks, UNDP), where 'adaptation' receives the most attention. For example, #COP18 is the hashtag for the 18th Conference of the Parties, which took place in Doha and is present in both 'adaptation' and 'mitigation' tweets. #UNFCCC is present in relation to 'mitigation.' A noteworthy top hashtag is #agriculture, a food-related issue, also present in both 'mitigation' and 'adaptation' tweets, but (again) with a larger occurrence in the 'adaptation' collection.[80]

The 'skepticism' and 'conflict' profiles both offer up distinct discursive spaces. The resonance of skepticism is dominated by actors that are, in fact, critical of climate change skepticism, rather than being skeptical themselves of human-induced change. Furthermore, it is striking how the top users are recognized throughout this space, as will become clear from the following example of the Twitter user named Skepticscience. @Skepticscience is the most-mentioned user for 'skepticism,' and with 2684 mentions is even the most-mentioned user across the board, outnumbering those for 'mitigation', 'adaptation' and 'conflict'.[81] The user is connected to skepticalscience.com, a website with the slogan 'getting skeptical about global warming skepticism', which is the top host in the 'skepticism' collection.[82] This underlines the importance of combining computational analysis with a qualitative close reading of the data, with attention to the actors and their content. A solely quantitative analysis, in this case, would have lead to misinterpretation of the results, concluding a strong presence of skepticism, where, in fact, criticism of skepticism resonates strongly here.

78　This is similar to working with top results in the Google Web Search engine; it follows the logic of the medium and the logic of working with the results most viewed (and clicked) by its users.

79　The tweets were checked for false positives by close reading the top tweets to see whether these indeed refer to climate change. The reason to focus on top tweets is that these are not only the most prominent according to the logic of the platform itself but (similar to search engine results that are high in the ranking) they are also the tweets with the most exposure and therefore are most viewed by users.

80　EMAPS, 'Profiling Adaptation and Its Place in Climate Change Debates With Twitter'.

81　As the data set contains retweets too, it could occur that a single message that is often retweeted skews the data heavily. Therefore, it is important to read the data closely to interpret the results.

82　SkepticalScience, 'http://www.skepticalscience.com/' is also prominent in the search engine case study in Chapter 5.

Scepticism	Mitigation	Adaptation	Conflict and violence
QUERY	QUERY	QUERY	QUERY
Skeptic	Mitigation	Adaptation	Conflict OR Violence

TWEETS ▣No Link ▣Link **USERS**
24% 30.560 76% 19.068

TWEETS **USERS**
40% 9.834 60% 6.742

TWEETS **USERS**
26% 31.533 72% 14.852

TWEETS **USERS**
58,1% 18.065 41,1% 10.818

TOP 10 HASHTAGS

Scepticism	Mitigation	Adaptation	Conflict and violence
climate [2674]	Climate [1234]	climate [5423]	ClimateChange [1995]
ClimateChange	climatechange	adaptation	climate
globalwarming	Mitigation	climatechange	violence
tcot	COP18	COP18	globalWarming
auspol	adaptation	agriculture	water
ClimateCrisis	flood	resilience	nra
p2	UNFCCC	CBA7	autism
agw	Agriculture	mitigation	conflict
science	Doha	climateadaptation	obesity
skeptics	auspol	africa	add

TOP 10 MENTIONED USERS

Scepticism	Mitigation	Adaptation	Conflict and violence
Skepticscience [2684]	Green_Register [347]	IIED [348]	LOLGOP [574]
dana1981	WorldBank	Green_register	MotherJones
Elonmusk	Cejarclimate	Guardian	CainTV
Bencubby	UN_ClimateTalks	Cejarclimate	THEHermanCain
Algore	RepJBridenstine	Alertnet	guardian
Huffpostgreen	FAOclimate	Weadapt1	CuestionMarque
Fxnscitech	SEIclimate	Seiclimate	FredZeppelin12
Michaelemann	CFigueres	Euenvironment	Greenpeace
Huffpostpol	UNESCO_AsiaPac	Worldresources	hrw
Climatedepot	UNDP	Joerogan	Oxfam

TOP 10 ACTIVE USERS

Scepticism	Mitigation	Adaptation	Conflict and violence
Dana1981 [222]	Ironsidewater [45]	Mklondeveloper [357]	Donbeeman [4155]
Drtucker	Kidbrightwillow	Maskafrika	Intipolitical
iluvco2	Seiclimate	Jogmans	Danielsbuk
Kernos501	Adbclimate	Seiclimate	Gcmcdrought
Sydnets	Qldaah	Weadapt1	Cuestionmarque
Octarediane	Kings_cambridge	Acclimatise	Stacydvandeveer
Michaelemann	Qwadja	Adbclimate	El_climate
Skepticsscience	Life_sciences	Nccorfian	Global_policy
Climatedepot	Strategikas	Apanadapt	Whohatesobama
4589roger	Stacydvandeveer	Alertnetclimate	Seedbomb4changea

TOP 10 HOSTS

Scepticism	Mitigation	Adaptation	Conflict and violence
skepticalscience.com	thegreenregister.com	trust.org	guardian.co.uk
huffingtonpost.com	ccafs.cgiar.org	guardian.co.uk	motherjones.com
foxnews.com	sciencealerts.com	mask-africa.com	theguardian.com
forbes.com	facebook.com	itunes.apple.com	bloomberg.com
thehill.com	climatechange.worldbank.org	grist.org	trust.org
washingtonpost.com	youtube.com	thegreenregister.com	wagingnonviolence.org
guardian.co.uk	ebc.net.au	ccafs.cgiar.org	thinkprogress.org
blog.algore.com	trust.org	weadapt.org	youtube.com
wattsupwiththat.com	triplepundit.com	sciencealerts.com	news.yahoo.com
youtube.com	thinkprogress.org	allafrica.com	sciencedaily.com

MOST LINKED URL

Scepticism	Mitigation	Adaptation	Conflict and violence
610	355	531	230
Peer-reviewed survey finds majority of scientists skeptical of global warming crisis	Climate change mitigation and adaptation equally critical for global food Security	Food Security through Investors and Government	Climate change's links to conflict draws UN attention

MOST RETWEET

Scepticism	Mitigation	Adaptation	Conflict and violence
503	70	367	550
RT @elonmusk: In reality 97% of scientists agree that we face serious human generated climate change http://t.co/soQCnJB61B	RT @RepJBridenstine: Pres. Obama is pledging billions of your hard-earned dollars for "international climate mitigation and adaption projects"	Alarming climate change effects on fl. ""managing the risks of extreme events and disasters to advance climate change adaptation"" is out...	RT @lolgop: fun fact: the same people who oppose doing anything about gun violence are also wrong on climate change the economy everything else.

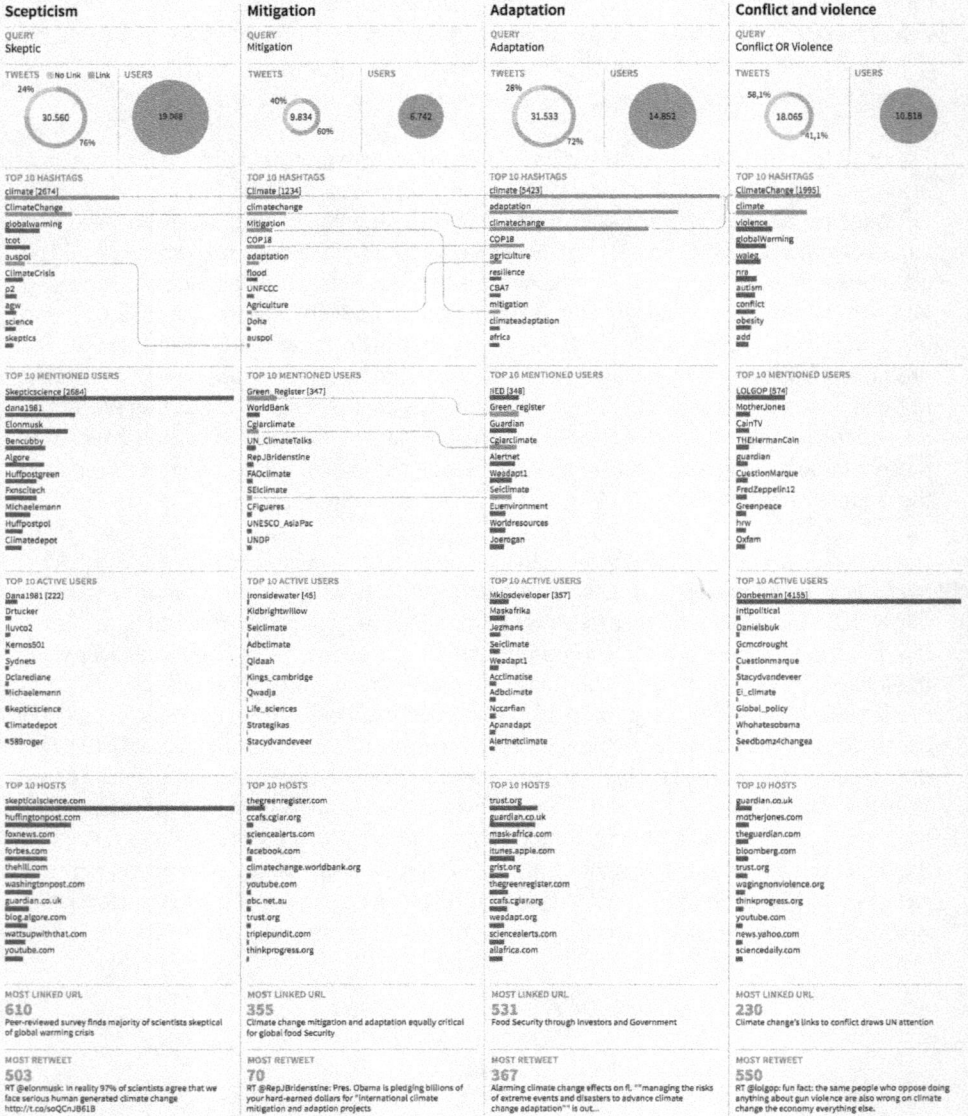

Figure 16: Profiling adaptation and its place in climate change debates with Twitter. This map shows profiles of four discursive areas within the climate debate: skepticism (with the query 'skeptic'), mitigation, adaptation, and conflict (with the query 'conflict' OR 'violence').[83]

83 EMAPS, 'Profiling Adaptation and Its Place in Climate Change Debates With Twitter', 2014, http://climaps.eu/#!/map/profiling-adaptation-and-its-place-in-climate-change-debates-with-twitter-I.

SkepticalScience's survey report *The Consensus Project*, which assessed over 12,000 peer-reviewed climate science papers for consensus on human-induced climate change, is the object that resonates most in this space, receiving 660 links and listings in the most retweeted message by the third most mentioned user, the entrepreneur @elonmusk: 'In reality 97% of scientists agree that we face serious human generated climate change http://t.co/soQCn-JB61B', which was retweeted 503 times.[84]

In conclusion, the profiles offer a view beyond the substance of the issue, to capture the actors present in this space as most active or most mentioned users, be they individuals or organizations. This way, the Twitter networked content analysis offers insights into the types of actors present in the debate and the intensity (and perhaps even interrelatedness) of their arguments and references. Again, we may productively ask: What kind of climate change debate does Twitter present? In the climate change adaptation and mitigation profiles, the most resonating users (mentioned) are international organizations working on the issue of food security. For example, the CGIAR (Research Program on Climate Change Agriculture and Food Security) ranks highly in both. Similarly, top users and hostnames are organizations, such as the Mask-Africa Food Security Program, in the case of adaptation. In the case of 'mitigation', when looking at the type of content that circulates best through the most shared URLs, the organization Green Register, a blog dedicated to environmental sustainability news and eco-friendly living tips, ranks highly. The top users actively engaging with the mitigation discourse are more diverse, and include companies, academics, and international organizations.

For the climate change skepticism profile, the top users are those skeptical of climate change skepticism, and the most-shared content acknowledges the human-made origins of climate change. News media rank highly, and famous protagonists of human-made global warming appear here, including Al Gore. The cluster also gives voice to journalists and entrepreneurs infamous for their skepticism. The interrelation between the scientific and the public debate is perhaps best captured by the Consensus Project. The Consensus Project takes an academically published scientometric analysis of climate consensus in climate science publications and publishes it in media campaigns stressing consensus on climate change. As the website theconsensusproject.org reads: 'Using peer-reviewed science, it plays an active role in debunking climate misinformation published across the spectrum of media, including TV, online, and print.' Its resonance is easily retrievable in Twitter, where it has performed as the most shared URL in the skepticism set. Relatedly, the study in the previous chapter showed the strong connections between skepticism and mass media, indicating the shift from a scientific to a public (and heavily mediated) debate. Similarly, conflict is associated with news media and public figures, for instance, radio show hosts (@hermancain), but also organizations with a humanitarian focus, such as Oxfam and Greenpeace, that address the humanitarian aspects of the environmental crisis.[85]

---

84   Cook et al., 'Quantifying the Consensus on Anthropogenic Global Warming in the Scientific Literature'
     See also Skeptical Science. 'The Consensus Project'.
85   EMAPS, 'Profiling Adaptation and Its Place in Climate Change Debates with Twitter'.

Having zoomed in on the most prominent issues and actors in climate change-related tweets, where adaptation and food security are leading issues, my analysis now takes a more exploratory approach (in the vein of Tukey's approach).[86] Hashtags included in the same tweets, for example, can form thematic clusters with a myriad of sub-issues illustrating the current state of climate action and adaptation. Co-hashtag analysis allows for the characterization of hashtags in terms of how they are networked associatively with other hashtags.[87] As discussed in the introduction, there are limitations to samples demarcated by hashtags. However, given the large dataset that I am attending to here — 4,771,135 tweets from 1,780,225 distinct users — this filtering by hashtag usage provides a sizeable yet manageable subset of sample data.[88]

Exploratory View: Co-hashtag Analysis of Climate Change Tweets

For the exploration of co-hashtags within the data set, we first visualized the thematic clusters that could be identified within the Twitter space, based on the 'modularity class' algorithm in Gephi, an algorithm that detects communities of densely connected nodes where the nodes belong to different communities more sparsely connected.[89] Considering the (still) large amount of data in the data set, we made use of the OpenOrd layout, a force-directed layout algorithm specifically designed to encourage clustering in densely connected, large-scale, undirected graphs.[90] As the nodes 'climate change' and 'global warming' generated the strongest results (as expected), we excluded them from the graph to render legible their sub-clusters. The resulting clusters were manually categorized into themes that captured the essence of the connected hashtags. We followed this with a close reading of the actual tweets involved to verify the themes.[91]

---

86   J.W. Tukey, 'Exploratory Data Analysis', 1977, http://xa.yimg.com/kq/groups/16412409/1159714453/
      name/exploratorydataanalysis.pdf
87   Gerlitz and Rieder, 'Mining One Percent of Twitter'.
88   For this case study, we took a dataset of tweets posted between 23 November 2012 and 23 November
      2013 containing the query [climate change OR global warming], consisting of 4,771,136 tweets from
      1,785,296 distinct users, using the tool TCAT. Borra and Rieder, 'Programmed Method'.
89   V.D. Blondel, J.-L. Guillaume, R. Lambiotte, and E. Lefebvre, 'Fast Unfolding of Communities in Large
      Networks', *Journal of Statistical Mechanics: Theory and Experiment* (2008): 2.
90   Shawn W. Martin, Michael Brown, Richard Klavans, and Kevin Boyak, 'OpenOrd: An Open-source
      Toolbox for Large Graph Layout', *Proceedings of the SPIE Visualisation and Data Analysis*, 2011,
      https://doi.org/10.1117/12.871402. The algorithm uses a so-called 'simulated annealing' schedule,
      with five different iterations in which several parameters are changed. In the first two stages, a strong
      edge-cutting strategy is employed: long connections between nodes are ignored, promoting clusters
      segregation and increasing at the same time the amount of white space in the layout.
91   This proved necessary to eliminate the noise of tweets unrelated to climate change, for instance, one
      discussing a positively changing *investment climate* in the Chinese real estate market.

An exploratory reading of the network graph in Figure 17 (p. 108-109) shows some aspects that have long been a subject of discussion where climate change emerges as a controversy object, both among scientists and the public. The network displays clusters focused on the two main approaches to dealing with the impacts of climate change: adaptation and mitigation. This is reflected in hashtags such as #adaptation, #preparedness, #mitigation, #resilience, #impacts and #naturaldisasters. More specific discussions of adaptation revolve around energy, solar power, and fossil fuels, explicated in hashtags revealing the need to take action to counteract the impact of environmental change, such as #gofossilfree, #fossilfools, #carbonfootprint, #cleantech, and #renewables.

Skeptical views on climate change are also addressed in the Twitter space. In this case, however, as seen in the profiles, the skepticism-related tweets mainly oppose climate skepticism. These users organize their content through hashtags such as #sciencesaysso, #actionaugust, #climatedenier, and #climatedenieraward. The hashtag #actionaugust refers to the August of 2013 when the Organizing for Action movement delivered unicorn-shaped 'climate denier awards' to congressional members skeptical of climate change, 'ignoring the overwhelming judgment of science'.[92] The prominence of actions against known climate change skeptics and their institutional networks shows that views acknowledging climate change still take a more central position in the Twitter co-hashtag network than those skeptical of its man-made production and mitigation.

Of specific interest, is the number and variation of thematic clusters of climate vulnerabilities and casualties that can be identified specifically through the networked content analysis of climate change in Twitter. Here, tweets clustered by hashtags range to include everything from vulnerable animals and habitats to victims of extreme weather events. The majority of vulnerability-related clusters is concerned with marine habitats and the vulnerability of the Arctic, as hashtags like 'northpole', 'antarctic', 'melting', 'overfishing', 'oceans' prevail. In particular, 'reefs' and 'antarctic' are named in this context as vulnerable spots, where 'polar bears', 'penguins', 'whales', 'trout' and 'sharks' seem to be the most prominent *issue animals* threatened with injury, death, and, ultimately, extinction.[93] The biodiversity cluster reflects the need for resilience towards climate change for 'birds', 'turtles', 'koalas', 'tigers', and 'butterflies', again pointing towards the vulnerability of habitats and species.[94]

---

92   K. Burkhart, 'Organizing for Action Delivers Unicorn Trophies to 135 Climate Deniers in Congress', *The Huffington Post*, 13 August 2013, http://www.huffingtonpost.com/2013/08/13/organizing-for-action-climate-deniers_n_3750126.html.

93   These clusters reflect indicators of habitat change as included in the DARA index, and ecosystem services as defined in the ND-Gain index, albeit mostly focused on animals rather than indicators of effects on human habitats. A number of countries are however, mentioned in the context of #drought and #rainfall, such as Haiti, Namibia, Malawi, Jemen, and Liberia.

94   See also the *Issue Animals* study by Niederer and Weltevrede. Digital Methods Initiative, 'Networked Content', 2008, https://digitalmethods.net/Digitalmethods/TheNetworkedContent. Rogers, *Digital Methods*.

Climate Change Tweets Co-Hashtag Clusters

Twitter set
global warming

Query
climate change
OR
global warming

Date range
23 November 2012
23 November 2013

Size
4.771.135 tweets from
1.780.225 distinct users

Layout
OpenOrd Layout

Hashtags

Clusters

Note
climate change and *global warming*
nodes have been filtered out.

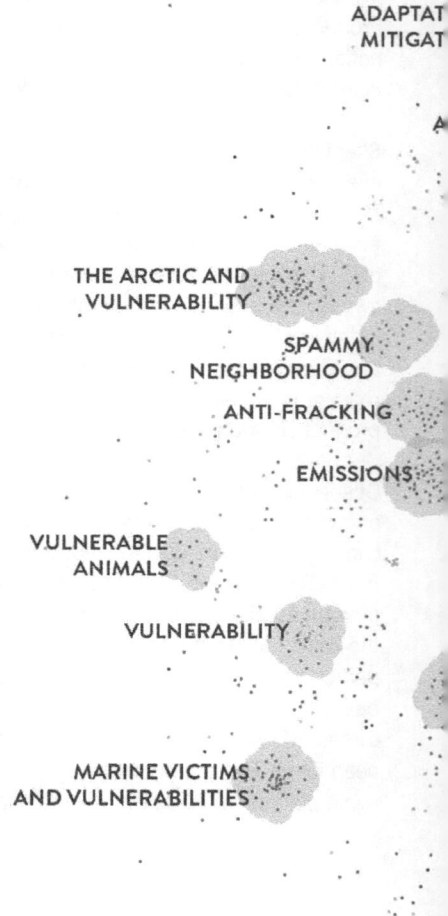

ADAPTAT
MITIGAT

A

THE ARCTIC AND
VULNERABILITY

SPAMMY
NEIGHBORHOOD

ANTI-FRACKING

EMISSIONS

VULNERABLE
ANIMALS

VULNERABILITY

MARINE VICTIMS
AND VULNERABILITIES

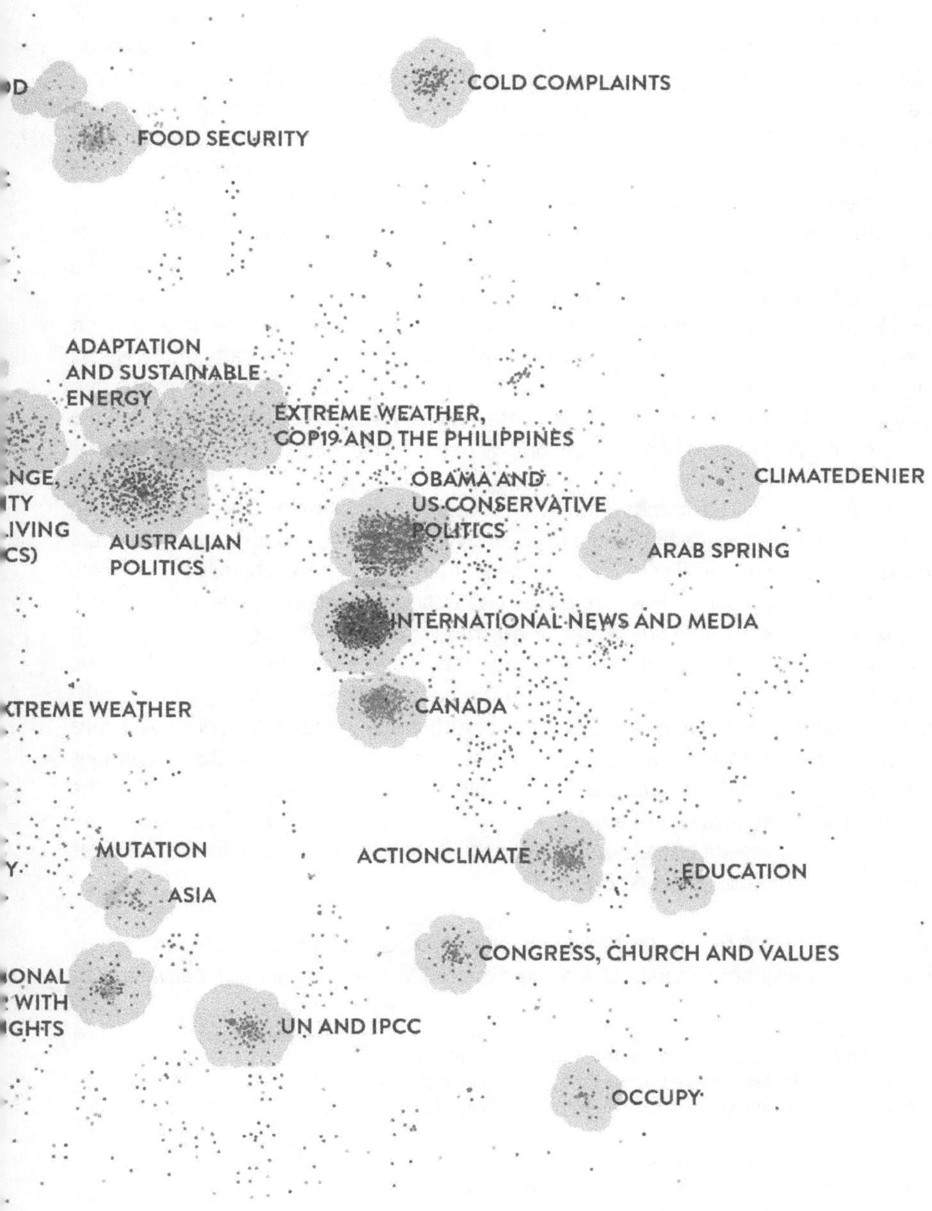

Figure 17: Climate change tweets co-hashtag cluster map. This network visualization shows thematic clusters in hashtag clusters within a set of climate change tweets.[95]

95   EMAPS, 'Climate Change Tweets Co-Hashtag Cluster Map', 2014, http://climaps.eu/#!/map/profiling-

Other clusters tend to focus on geographical regions such as Australia, Canada, and the US, albeit mainly in terms of climate change or global warming as an important topic in the national political agenda. Particularly dense clusters reflect specific, localized takes on the issue of climate change, centered on political events. One of these clusters focuses on Obama and conservative U.S. politics, with hashtags such as #obamacare, #obamaisnotsatan #inauguration2013 mentioned alongside #climategate #badscience and #globalwarminghoax. In his inaugural speech in 2013, Obama emphasized the need to respond to climate change as a threat to future generations. He further stressed the urgency of action when he argued: 'some may still deny the overwhelming judgment of science, but none can avoid the devastating impact of raging fires and crippling drought, and more powerful storms'.[96] A Canada-themed cluster reflects both the political events in Vancouver and the climate hearings in Saskatoon. The events in Vancouver revolved around the British Columbian Green Party in the weeks leading up to the elections, addressing the need to reduce greenhouse gas emissions and opposing oil pipeline expansions.[97] Canada conservatives, on the other hand, supported the expansion of oil pipelines, spurring a debate on Twitter regarding the facts of climate change.[98]

In November 2013, the Saskatchewan citizens' hearings on climate change also gained prominence on Twitter through the hashtag 'climatesk'. The hearings included a two-day event with presentations on the realities of climate change from scientists, teachers, newcomers to Canada, and affected groups, allowing the voiceless victims of climate change to be heard.[99] A third political cluster includes hashtags on Australian politics, reflected in hashtags such as #melbourne, #auspol, #ausvotes, and #carbontax. The debate there revolved around Tony Abbott, prime minister of Australia since 2013, and his statements of September 2013 announcing that he would not increase funding for further carbon tax reductions if Australia missed its emission reductions target.[100] These clusters thus seem to detail the discussions on statements made about climate change following specific political events, as well as the political views of those involved in elections around the world. Each identifies clear opportunities for scholarly research that uses Twitter as a 'source of current and topical news' as proposed by Phelan, McCarthy, and Smyth.[101]

Clusters that are formed by hashtags related to official sources (UN and IPCC), climate activism and everyday weather remarks additionally express that many, if not most, conversations

adaptation-and-its-place-in-climate-change-debates-with-twitter-ii.
96   Stevenson and Broder, 'Climate Change Prominent in Obama's Inaugural Address'.
97   J. MacNab, 'Will Climate Be a Winner in British Columbia's Election?' *Pembina Institute*, 2013, http://
     www.pembina.org/blog/724.
98   C. Cattaneo, 'As B.C. Election Looms, Both NDP and Liberals Take Hard Line on Oil Pipelines',
     Financial Post, 2013, http://business.financialpost.com/2013/05/06/
     bc-election-oil-sands/?__lsa=d52e-3289.
99   'Saskatchewan Citizens' Hearings on Climate Change', 2014, http://skclimatehearings.org/.
100 L. Taylor, 'Rudd Accuses Abbott of Abandoning Australia's Alimate Commitments, *The Guardian*, 13
     September 2013, http://www.theguardian.com/world/2013/sep/03/rudd-abbott-
     abandoning-climate-commitments.
101 O. Phelan, K. McCarthy, and B. Smyth, 'Using Twitter to Recommend Real-time Topical News', in
     Proceedings of the Third ACM Conference on Recommender Systems, 2009, pp. 385-388, http://
     dl.acm.org/citation.cfm?doid=1639714.1639794.

on Twitter emerge around particular (current) events and other real-time experiences.[102] This endorses the aforementioned scientific literature on Twitter as a medium for real-time and happening content. Lastly, the network further displays resonance of the previously profiled climate change discourse of conflict. These very small clusters of hashtags focus on the Arab Spring in particular, and hashtags such as #arabspring, #libya, #syria, #egypt, #morsi, #iran, and #drones appear here. A close reading of these tweets reveals that in part, the relation between climate change and conflict is popularly recognized on Twitter, with re-tweets from news articles on the issue, but these being also skeptically assessed, in tweets similar to the following: 'Syria conflict is not caused by drought. Its more to do with a bad mix of Religion and politics just like everywhere else.'[103]

Taken as a group then, these Twitter clusters provide a rich snapshot of the state of the climate debate, indeed work as a kind of 'awareness system,' to speak with Hermida, that gives voice to the different voices and actors active in this realm, and reveals the intertwinement of the news and other mass media content with the platform's content.[104] Twitter does not produce 'climate science' but instead puts scientific research into circulation, while also enabling up close, located and platform-literate engagements that assess the resonance of climate change adaptation and indicators of vulnerability within the broader online discussion of climate change. It should be noted, however, that this awareness system is only accessible by combining computational analysis with a qualitative close reading of the data, with attention to the actors and their content. As mentioned, Twitter's entanglement with news journalism and mass media should be kept in mind here, as Twitter *is* news, amplifies news, and is a channel for news distribution.

Conclusions

On a methodological level, we may conclude that networked content analysis applied to Twitter content entails working with the logic of the platform and recognizing the socio-technical structures of its content. By attending to the natively digital elements of this platform, it becomes possible to assess how content is networked and circulated. In the case study presented here, I compare the resonance of three different climate change discourses. After demarcating a specific set of issue-related tweets, I query the set for the resonance of recognized keywords to create 'keyword profiles'. Important to note is that, against the rise of 'big data' pattern recognition, a close reading of the data proved necessary to correctly interpret the found data and further filter the data to improve its relevance.

---

102 As expected, mundane climate change 'updates' also find their way into the climate change Twitter
     network, with complaining tweets about the cold weather that include such hashtags as
     #cold and #freezing.
103 The small clusters include tweets recognizing the connection, such as '#arabspring caused by
     #history's most underrated force: #climate change'. The discussion also includes skepticism towards
     this connection between climate change and conflict, as with the example given in the text and with this
     tweet: 'seriously global warming err climate change caused Syria? Unreal'.
104 Hermida, 'Twittering News'.

The keyword profiles offer zoom-in views on particular discourses within the broader issue of climate change. Here, looking at skepticism, mitigation, adaptation, and violence, the profiles enable a comparative view, and it becomes clear that mitigation and adaptation are very proximate issues, in terms of argumentation and actors, with most actors involved in the Twitter space being organizations in the field of food security. The UN and its initiatives dominate both discourses. Adaptation has now surpassed mitigation in terms of resonance; thus, here also in Twitter, the so-called adaptation turn discussed in the introduction of this chapter has taken place. Skepticism and conflict have distinct profiles, where a focus on 'skepticism' as a keyword brings up (perhaps counter to expectations) much criticism of climate change skepticism. This is mainly organized around the actor Skepticalscience (both as a user named @Skepticalscience and as a website host) and major news sources, and makes an important point for the close reading of data rather than favoring coarser pattern recognition. In the profile of conflict, news media, and media personalities resonate strongly, as do humanitarian NGOs.[105]

The exploratory analysis of the climate change Twitter hashtag network shows us that climate change as a controversy object appears through, or is a sum of, a multiplicity of sub-issues including skepticism, mitigation, adaptation, vulnerability, and conflict. Just as the comparison of different Indices' rankings revealed conflicting vulnerabilities, the Twitter hashtag network also points to different metrics of vulnerability.[106] In contrast to how vulnerability indices organize and rank vulnerability by country listings, it appears that the 'nation' is not the key entity we are tracking in relation to vulnerabilities registering on Twitter. Rather, the objects of vulnerability and injury that are put forward on Twitter are mostly animal species and habitats (which, needless to say, are categories of 'actors' entirely oblivious to legally drawn borders). As discussed in the Introduction, animals are mediagenic issue actors, and some are more mediagenic than others. Nevertheless, it is important not to overstate mediagenic power and take the prominence of animal species and habitats seriously as they appear. I would like to argue that such a framing of risk in terms of present and future risked species, ecologies, and systems provides a possible Beckian extension to the study of Bruns and Liang, who described Twitter as a powerful channel for crisis management after the fact of a natural disaster.

The networked content analysis of hashtag clusters that are dedicated to sub-issues, casualties and events can be read as a time slice presenting the status quo of climate change, one that is not merely stating 'what's happening' but rather serves as a progress report on an issue, in this case both addressing where we are with climate change adaptation and what is at stake.

---

105 The profile, however, also shows that this specific discourse also seems to have been hijacked by a single user trying to widen the issue by connecting it to medical conditions such as obesity.
106 While I focus on Twitter analysis here, the EMAPS study also analyzed the prominence of adaptation and the other discourses on the web as accessed through Google. Querying the keywords 'skepticism,' 'mitigation,' 'adaptation' and 'conflict' in the top Google results for the query 'climate change OR global warming,' we also found 'adaptation' to be the most widely present keyword in Google top results about climate change.
      See also EMAPS, 'Reading the State of Climate Change From the Web: Top Google Results', 2014, http://climaps.eu/#!/map/profiling-adaptation-and-its-place-in-climate-change-debates-with-twitter-ii.

6. CONCLUSION

In this book, I argue that the analysis of digital media content needs an approach that takes into account the specificities of how platforms and engines serve, format, redistribute, and essentially co-produce content. These specificities are what I refer to as the *technicity of content*. The foundational work that established the field of content analysis, developed within communication science, paved the way for the analysis of (large) bodies of text for features or (recurring) themes, in order to identify 'cultural indicators' or make other inferences about societal trends and issues.[1] While content analysis has seen a tremendous uptake across scientific disciplines, the application of these methods to *networked* web content has presented an ongoing challenge for researchers of various scholarly disciplines. Therefore, I propose to improve the adaptability and fit of content analysis to networked content through a range of digital methods and tools that I show to be conducive to the task. The work of Klaus Krippendorff, a major proponent and methodological innovator of content analysis as a field of media research, is a key driver of my own development of what I name and develop here as *networked content analysis*. As content analysis has been inclusive of content (in all shapes and forms) *and its context* since its early beginnings, its methods only need to be amended to suit the digital era and deal explicitly with the technicity of networked content.[2] I propose to utilize controversy mapping and digital methods to do so, building on these methods' respective actor-/issue-centricity and medium-specificity.

In this book, I develop these research techniques through the analysis of the climate change controversy, an ongoing debate that takes place across scientific disciplines and into the public realm, across platforms, sources, and studies, from the first international climate skeptics conference of 2008 all the way to 2015. When I started this research, the climate controversy was publicly understood as historic but hard to historicize, as it was being lived in real-time. It also experienced an upswing in debate temperature once skeptics began organizing themselves in these annual conferences, and as several publications rose in response to debunk their status and unveil skeptics' entanglement with industry funding, especially tobacco and oil industries. The case studies in this book end in 2015, when 198 countries signed the 'Paris Agreement' to cut back on CO_2 emissions in a joint effort to turn the tide of climate change. As we now know, a year after that milestone, Donald Trump, announced the United States' withdrawal from the Paris Agreement, which has lead to a surge in climate discussions and climate activism in social media and in the streets.

With this book, I do not aim to contribute to climate *science*, which is well outside of my area of expertise, but instead to offer a contribution to the study of online content by developing a networked content analysis of the climate *controversy* as it is specifically formatted and transformed by platforms and actors. The study accordingly follows the climate debate in science, as well as on the web (and Google Web Search), in Wikipedia and in Twitter, and analyzes how content is networked there, in order to propose adaptive and sensitive research techniques appropriate to networked content analysis.

---

1   Gerbner, 'Toward "Cultural Indicators"'.
2   Krippendorff, *Content Analysis*, 2004.

These research techniques draw from existing approaches and methods developed to study controversies and their actors (controversy analysis) and social and cultural issues with the web (digital methods). Controversy analysis gives direction to the study of controversy without the translation of actor language into preset categories listed in a codebook. On the contrary, it makes a case for descriptive research and advises researchers to launch their inquiries 'in medias res' and describe what they see.[3] There is no single specific protocol, toolkit or methodological framework for controversy analysis, but there are 'commandments', publications, and an educational program at Sciences Po in Paris that provide many guidelines for the operationalization of the mapping of controversies.[4] In my case studies, which analyze the climate debate on the web, Wikipedia and Twitter, this leads me to describing the group formations (on the web) of climate actors, following actors across networked and forked articles about climate change and related topics (Wikipedia), and exploring and describing climate change co-hashtag networks in Twitter. Digital methods are developed at the educational program at the University of Amsterdam in close kinship to controversy mapping and provide concrete tools and methods for the study of web-based dynamics of social and cultural issues. Similar to controversy analysis and content analysis, digital methods put forward non-intrusive methods, views, aspirations, and affiliations of issues and their actors, collecting data from websites and social media activity.

The differences that I outline between content analysis as it was incepted, controversy mapping, and digital methods, all with certain limitations, are reiterated throughout the chapters of this book. Krippendorff's robust articulation of content analysis for a prior media age conceptually acknowledges but strains methodologically and tool-wise, to grapple with the networked qualities of online content, where issues, debates, and actors may spread out or recur across platforms and other carriers. The addition of controversy analysis offers a research outlook to follow actors and describe the many viewpoints and stakeholders present in a debate while being under-attentive to operationalizing this with regard to networked content by offering mapping methodologies that deal with digital media content. This is where digital methods come in, which offer tools to capture and analyze an issue through networked web content that otherwise is not available in content-driven communication research. My main contribution here lies in the combination of these approaches that makes possible the content analysis of networked content.

Networked content analysis as content analysis that is amended to suit online networked content enables a researcher to jump in the middle of a controversy, follow actors and describe these actors' viewpoints in their own words, employing digital methods to capture and analyze the substance of debates across platforms. My proposition here is similar to Susan Herring's 2010 study in so far as I also am interested in a widening of the paradigm of content analysis with methods from adjoining scientific disciplines. However, while Herring regards content as contained in media documents, I argue that such a separation between content and its carrier cannot hold with networked content. Furthermore, tracing the discipline of content analysis backward, I note that such a division between content and form or carrier is quite

---

3   Latour, *Reassembling the Social*, 27.
4   Venturini, *Diving in Magma*.

antithetical to the way that Content Analysis was originally conceived by Krippendorf. Nevertheless, Krippendorf's formulations are pre-web. Understanding the technicities of the platforms that serve and co-produce content today entails studying platforms' characteristics and identifying the queries or tools that are necessary to demarcate and analyze networked content. Studying platforms as socio-technical systems is of the utmost importance, as they are 'increasingly embedded in our societies'.[5] In this book, I develop such a socio-technological perspective on the controversy surrounding climate change as presented and debated on the web.

Krippendorff, in his foundational work, stressed that it is one's definition of what content is and how that is delimited that leads to specific kinds of analytical results. As we have seen with the analysis of web content in the various case studies, it is indeed this very refinement of defining (the materiality of) content that, with the recognition of the technology as an active material agent and part of content leads to a specific demarcation of content online. As I argue in Chapter 2, the definition and demarcation of content have never been so straightforward in the case of offline materials, and changing technologies have further complicated these matters. The digitization also of analog content has changed the nature of materials already, raising new questions regarding the inclusion of features and formatting in the analysis. With hyperlinks, content became networked, and thus, it became harder again to demarcate and to establish where so-called content ends. Search engines brought about new ways of presenting and ranking data, and platformization gives further shape to the far-stretching entanglement of social media with other web content.[6] Network Content Analysis aims to be adaptive to the specific technicities of platform content; therefore, I approach the climate debate on each platform with platform-specific ways to define and delineate the corpus to analyze. In my case study of the web, I demarcate sources, for instance by taking the speakers list from an international climate skepticism conference and looking up their respective websites to use for further research, and by taking the top results for the query' climate change' in various languages to measure the resonance of prominent actors. In the Wikipedia chapter, I discuss a study of climate change-related articles in which the demarcation occurs by taking those articles that are reciprocally linked from the climate change article.[7] In the Twitter study, I demarcate Tweets by a query that includes tweets containing climate change or global warming.

The inclusion of web content's technicity into the idea of content itself then leads to analyses that make use of and deal analytically with, these technical agents. The collection and analysis of web content that follows the specificities of each platform, and operationalizes the specific technicities at play, will lead to more precise analysis, one that is sensitive to the networked nature and dynamical movement of online content. I realign my work with Krippendorff's inceptive call to keep the content together with its carrier (or context), and accordingly propose that in Networked Content Analysis, researchers include not only the carrier (e.g. the search engine result, the Wikipedia article, the tweet) but also the technicity thereof (e.g. the ranking of the search results, the editing history and content robots of the Wikipedia article,

5   Lazer et al. 'The Parable of Google Flu', 1205.
6   Helmond, *The Web as Platform*.
7   'Reciprocal linking' here means both linking to and receiving a link from the article on climate change.

and the hashtags and retweets networking a collection of tweets) as part of their analytical approaches. This book offers up new ways forward for content analysis approaches, methods, and techniques that are suitable for the study of online networked content. Rehabilitating the inceptive work of Krippendorff, the contemporary web-literate approach to networked content analysis that I demonstrate here remains open to all kinds of content and includes contents' technicity as part of its research method.

Applications of Networked Content Analysis I: The Web

In the first case study, I approach the climate controversy by assessing the positions and affil-iations of its actors, starting at the time of the first international skeptics conference organized by the Heartland Institute in 2008. I analyze the networks of climate debate actors (including the conference's keynote speakers) using scientometric analysis, as well as techniques that I propose as being fruitful for networked content analysis. Namely, these are hyperlink analysis and search engine results resonance analysis, which I use to research the place and status of climate skepticism within both climate science and the climate debate as it takes place beyond the scientific literature. I approach the *networkedness* of content through hyperlinks to analyze networks of association. Subsequently, by using Google Web Search (to many a dominant entry point to the web) to demarcate top sources for the query of climate change, the case study zooms in on climate change actors and their prominence, as identified by the search engine. Here, I ask how the technical logic of search might be used to measure such prominence of actors in a specific issue, in this case, looking at the resonance of climate change scientists (both skeptical and non-skeptical) within a demarcated set of websites. I zoom in on a particularly heated moment of the debate in the Dutch context, immediately following a publication on the scientific consensus regarding climate change, published by the Royal Academy of Sciences of the Netherlands.[8]

Hyperlink analysis shows a distinct profile for the Dutch skeptics, who strongly associate themselves with the Anglo-American network that gathers at the Heartland conferences. Meanwhile, non-skeptical scientists, those 'climate-concerned', if you will, show a much more heterogeneous network, with links to science, government, UN, Worldbank, and mainstream media. Resonance analysis, in this case, shows less strong differences between skeptical and non-skeptical scientists, with both sets of actors resonating across sources, and coming in at the top and bottom of the search results. There are no sources that mention only the small sample of non-skeptical scientists without also mentioning the skeptics, but two sources that only give attention to skeptics. Lastly, through a close reading of the climate skeptics' websites, I find that their 'skeptical' delegitimizing campaigns extend to coverage of topics well outside of the realm of climate science (e.g., the health dangers of second-hand smoke). Paired together here, traditional scientometrics and techniques of networked content anal-ysis offer a fine-grained picture of the status, group formation, and issue commitments of climate change skeptics (compared to non-skeptical actors). While with scientometrics alone I have not been able to identify the skeptics as entirely separate from climate science as an academic field or inter-discipline, with Networked Content Analysis I have found divergent

8   KNAW, *Klimaatverandering, Wetenschap en Debat*.

networking behavior, as well as the aforementioned *related* issues, which qualified them more as *professional skeptics* rather than professional climate experts.

The main challenge that web-based media presents to traditional content analysis is that web content is networked, for instance, by hyperlinks. Another way that it is linked and processed is through social media buttons, which pull the content of websites into various platforms.[9] Furthermore, the fact that web content is often accessed through search engines such as Google Web Search, which rank and suggest content through undisclosed and ever-evolving algorithms, is just as problematic.[10] Asking subsequently what kind of climate change debate the web puts forward through such search technicity, I would conclude from this case study that it demonstrates actor alignment in networks of affinity, association, critique (as the skeptics linking to their main object of criticism: IPCC), and aspiration (e.g. in the case study, Dutch skeptics are hyperlinking to their Anglo-American colleagues, who do not link back). Resonance analysis reveals on one level the sources present in the top results of a query, but also the mention such sources make of specific keywords, or, in this case, actors. A close reading of these actors' websites establishes the image of their professional skepticism, problematizing, and delegitimizing the apparent professionalism of their commitment to climate change as an issue.

Applications of Networked Content Analysis II: Wikipedia

My second case study focuses on the climate debate on Wikipedia, the most well known go-to online and free reference system on the web. Characterizing the project as a socio-technical platform for knowledge production in the encyclopedic format, in this chapter, I discuss the dependency of the platform, its various user groups, and its content, on the (underlying) technicity of Wikipedia. In its status as an encyclopedia, it seems initially counterintuitive to think of Wikipedia as a space to study controversy. However, due to the way Wikipedia content is networked, designed, and managed, the platform has emerged to be recognized as a unique site for controversy mapping; this is because an online encyclopedic project is ever exposed as being 'in the making'. After discussing in detail the technicities and protocols of the Wikipedia project, I present two studies that each offered a close reading of a controversy that takes place behind the scenes of Wikipedia articles. I choose these specific studies and approaches in order to make a case for a networked content analysis that uses the (ever-evolving) technicity of this ubiquitous platform of Wikipedia in the analysis of a particularly contested and major controversial topic.

Subsequently, the networked content analysis of the climate change debate on Wikipedia by Gerlitz and Stevenson deploys the hyperlinks between articles on the topic of climate change to demarcate a network of related articles, which allows for the study of the composition of its editors (including active bots) as well as editing activity over time. Here, networked content analysis permits a historical reconstruction of the debate, and indicated generic Wikipedia editing trends over time, but also recognizes issue attention cycles, where 'new news' around

---

9   See also Helmond, *The Web as Platform*.
10   McMillan, 'The Microscope and the Moving Target'.

the controversy or debate has the effect of spiking Wikipedia activity across specific pages.[11] Lastly, heat maps may be used to signal significant moments in Wikipedia's 'management' of the issue of global warming, as I discuss extensively in the chapter. Here, the technicity of the platform formats content in a way that both its historicity and conditions of production (e.g., the talk pages) become visible to both users and researchers.

I am attentive here also to the periodization of research on Wikipedia, and its uptake by researchers as a tool. The first generation of scholarly Wikipedia research has focused main-ly on the platform's capacities for crowdsourcing knowledge production, as well as on the reliability of its co-produced content. I argue for more attention to the machinery that facil-itates and formats this knowledge production. While traditional content analysis reaches its limits to struggle with the omnipresence of technical agents in the wiki-platform of Wikipedia, networked content analysis provides means to properly assess Wikipedia's content, across articles and language versions. It can, of course, also still be used to compare web-based encyclopedia content to more static encyclopedia projects. All such potential research queries demand appropriate research frameworks and tools capable of capturing how Wikipedia is socio-technically modulated towards reliability and consensus over time.

Applications of Networked Content Analysis III: Twitter

In the final case study of Chapter 5, I study the state of the climate change debate in Twitter, which I commence by assessing the logic of this platform and how it networks and circulates its content. Here, I demarcate a set of climate change-related tweets using a tool called TCAT, and query the set for the resonance of recognized keywords from various discourses within the climate change debate. I present the results of this so-called resonance analysis as dis-course-specific *keyword profiles,* which allows for zooming in on the main actors and the main content circulating within this subset, providing insight in the different phases of the climate change debate. Importantly, and counter to practices of pattern recognition, a close reading of the data proves necessary to filter the collected data further, towards improved relevance.

The early applications of traditional content analysis discussed in Chapter 2 stem from the pre-platform era. Thinking back to the warning issued by McMillan to researchers wanting to use search engines, we can imagine the hesitation to work with APIs, and the differences between free APIs (offering limited amounts of data) and real-time full access to data (as opposed to for instance Twitter's 'Firehose' API), which often comes with a price tag. The main methodological contribution of this chapter is its development of a means to perform resonance analysis, where the demarcation of content (based on the literature of input from subject-matter experts) provides a sample in which the resonance of actors or keywords can be mapped. Similarly, the demarcation of tweets visualized through hashtag clusters allows

---

11   These general trends include an overall increase of editing interventions over time, a relative decrease
      in activity in the months of June and December, and the existence of an incubation period between an
      article's creation and its maturation, where after initial editing and a period of inactivity are followed by
      more regular editing.

for a descriptive and exploratory analysis of the debate around climate vulnerability.[12][13] The hashtag cluster network, I argue, could be read as a time slice, presenting the status quo of an issue or debate. In the case of climate change, this time slice does not merely state 'what's happening' but rather serves as a progress report or awareness system, addressing present challenges of climate change adaptation and what is at stake. The keyword profiles, on the other hand, enable a comparative view, which gives insight into how the discourse has shifted from mitigation to adaptation, confirming the 'adaptation turn', which has been declared in different realms. Furthermore, these methods enable a close-up study of the actors at the level of these distinct discourses. In this way, the Twitter study thus also underlines the persistent mutual interrelation between news media and platforms, whereby the platforms may produce news or act as a channel of distribution and amplification of content, sources, and actors, which I will reflect upon further in this conclusion.

Five Key Points

In this book, I discuss different research techniques that I propose together as an integral starting point for a practice of Networked Content Analysis. Some of these methods pre-exist my use of them for this purpose, while others are methodologically amended tools and techniques of digital methods. I would like to rehearse five key points, which establish the need for such techniques. Firstly, the main goal of this book is to develop an adaptive toolkit able to deal with the fact that different web platforms and engines serve content with different technicities. As each platform or engine has its own technicity and therefore requires specific methods and analytical tools, I try to stay true to the strengths of traditional Content Analysis for the humanities and social research — the non-intrusiveness of the method, the inclusion of content in all its shapes and forms, and the attention to the context of content — while further developing techniques that better adapt to the specificities of networked content.

Secondly, I find it important to emphasize also in my conclusion that content currently exists in and through the platforms and engines that produce it, which means a clean separation of content from its carrier is no longer feasible. It is now impossible or, at least, inadvisable to regard a Wikipedia article as entirely separate from its publicly available production process. Who were the authors? Were there bots involved? What is being presented as related articles? Which sub-topics (of an entry on Wikipedia) have become their own dedicated articles? Which were forked as a means of controversy management? Answers to these questions are likely to be of great interest and utility to those invested in Content Analysis in a networked era, and to anyone embarking on the mapping of a contemporary debate. Krippendorff has laid the groundwork for such analysis, well prior to content analysis having to deal with online content.

A third point I want to underline is that networked content also *folds in* traditional media content. Television news is published online, discussed in websites; news reports and images populate search engine results, lead to the creation of Wikipedia articles, or are linked to by tweets and amplified by retweets. This leads to the entanglement of news (and other mass)

12   Savage, 'Contemporary Sociology and the Challenge of Descriptive Assemblage'.
13   Tukey, 'Exploratory Data Analysis'.

media content, more traditional objects of study of content analysis, and networked content, the object of study in networked content analysis. The entangled nature of any media or content relation is where the focus and benefits of networked content analysis lie. In the Wikipedia study (of Chapter 4), I mention how news events tend to cause heightened editing activity in related articles. In the Twitter case (of Chapter 5), I discuss how Twitter as a micro-blogging platform could be approached through more conventional news cycle analyses but also through 'meme-tracking'.[14] In the latter mode, Twitter as a micro-blog could then be seen as highly responsive to or even parasitical or imploding of conventional news 'sites', echoing and amplifying news snippets by tweeting and retweeting. Further, as Twitter is often moving information faster than the news, Twitter content, in some cases, *is* news. Of course, for these reasons, Twitter is a popular medium for professional journalists. They bind tweets to their story, and when their work has been published, they may tweet a link to that article, using it as a channel for the distribution of their own work. As news and mass media sources strive to make their content 'platform-ready', a term by Helmond, the entanglement of news, other mass media content, and new platforms have entered the next level.[15] Networked content analysis proposes to take this entanglement as a given and demarcate content through the logic of the platform (as developed in digital methods) and thus follow the actors across sources (as key to controversy analysis). The rise of digital media does not mean the end of traditional mass media, but its reconfiguration as part of online networked content.

Fourthly, and more conceptually, I would like to propose that when studying the climate change debate through online content, we may regard the different platforms as different *windows* on the debate. Rather than asking 'What does Twitter say about the controversy,' or critically asking 'Who is on Twitter these days, anyway?' or 'Who uses hashtags?' we may productively ask: 'What *kind of* climate change debate does Twitter present?' 'And how does this compare and relate to the climate change debate as presented by Wikipedia (for example)?' In the climate change case studies in this book, the web presents a climate debate maintained by *professional skeptics* with distinct networking behavior and related issues and specific controversy objects. Wikipedia offers a view on a successfully forked issue, where the debate had been taken out of the main article, and the skeptical editors stayed true to the debate itself, migrating along to the new 'debate-article' established to address the controversy. Twitter presents a progress report of climate change adaptation, attentive to the landscapes and animal species endangered by climate change. In these ways, considering social media platforms as *windows on an issue* is also productive for creating a better understanding of the cultures of use of such platforms.

A fifth point worth mentioning is that while Wikipedia offers public views on its technicity, the other platforms studied in this book do not. Google Web Search, through its terms of service, does not allow for the use of its search engine for anything other than search. So repurposing the engine as a research device (as discussed in detail by Weltevrede) goes against its rules and regulations.[16] Twitter has various APIs; however, on an interface level, Twitter discloses

---

14   Leskovec et al. 'Meme-tracking and the Dynamics of the News Cycle'.
15   Helmond, *The Web as Platform*.
16   Weltevrede, *Repurposing Digital Methods*.

its mechanisms of ranking and prioritizing content (and neither does Google or any social media platform). This point was central to a critical project titled *The People's Dashboard*, which I developed together with Esther Weltevrede, Erik Borra, and others in 2015, and find relevant to mention briefly here.[17] The People's Dashboard is a social media platform plugin that visualizes the entanglement of content and users with the platform and its technicity. The dashboard is intended to be a critical layer on top of six different social media platforms: YouTube, Facebook, Twitter, LastFM, LinkedIn, and Instagram in order to discover and highlight 'people's content' as a layer on top of the interface. The plugin, which currently works for the interface of Facebook, color-codes interfaces of social media platforms according to whether it presents content *of the people*, or *of the platform* (Figure 18, p. 122). The project tries to increase understanding of what is actually *social* on social media nowadays. For researchers, such an understanding stresses the necessity to regard technicity as omnipresent, and make explicit how it is dealt with. This idea is recognized by scholars working with networked content such as Marres and Moats, who, in an STS-tradition, call for a symmetrical approach to the study of controversies with social media content, in which there is as much attention to 'media-technological dynamics' as there is to 'issue dynamics'.[18] Networked Content Analysis has a slightly different approach, as it proposes to include technicity by, straightforwardly, taking the *networkedness of content* into account. In the various case studies, I describe how platforms network content differently, and — as stressed in the first point — how this calls for an adaptive approach to the analysis of networked content, which is amendable to suit the technicity of a platform. Making technicity explicit in this way is comparative to the functionality of the People's Dashboard, as it offers a view on the entanglement of user content with the platform.

I would like to conclude here that networked content analysis remains true to its roots in content analysis as an unobtrusive method, while adapting to the web through medium-specific digital methods and taking on the research outlook of controversy mapping as a means by which actors may be followed, viewpoints traced, and presumptions left behind, in order to capture the richness and specificities of actor language. As such, it combines the adaptability and medium-specificity of digital methods and the richness of controversy mapping with the rigor of content analysis. Networked content analysis, as proposed through these kinds of imbrications, will give renewed significance to modes and methods of content analysis appropriate in and for the digital era.

---

17   *The People's Dashboard* is described extensively on the wiki project page, and the team members are listed in the 'Acknowledgements of collaborative work' section. Digital Methods Initiative, 'The People's Dashboard', 2015, https://wiki.digitalmethods.net/Dmi/PeoplesDashboard.
18   Marres and Moats, 'Mapping Controversies with Social Media'.

Figure 18: The People's Dashboard. This mockup of the People's Dashboard was developed during the Digital Methods Winter School of 2015, as a critical layer on top of the interfaces of dominant social media platform interfaces, revealing content of the people and of the platform. A third category is mixed content, indicating that people's content has been re-ordered or repurposed (e.g. Facebook News feed or birthday notifications). The plugin works with Facebook and is available on Github: http://bit.ly/peoplesdashboard.[19]

Technicities in Need of Attention?

In this book, I discuss the technicity of web search and interlinked websites, Wikipedia articles, and tweets. Of course, I have encountered many *technicities* beyond these that I did not discuss. Furthermore, there are many other platforms that could be studied in a networked content analysis of the climate debate. One could analyze climate activism in a large social media platform like Facebook, or a smaller image-based platform such as Pinterest, or study websites of climate change initiatives in a specific geographic region. For each platform, it is, in any case, crucial to ask questions that take into account the technicity of networked content: how is content networked in the platform, and what kind of issue does the platform present?

Perhaps one technicity that remained especially under-discussed and under-visualized in the maps is geo-location. And it is this question of place as an important technical aspect of networked content, which brings me to address the relationship between my proposed approaches and the directions and implications it has for future research. In the case studies, I map the major controversy of climate change, not by visualizing viewpoints on a traditional geographical map, but by tracing actors and (sub-) issues across online platforms. I discuss how platforms and engines such as Google Web Search enable researchers to focus on a

---

19   Digital Methods Initiative, 'The People's Dashboard'.

national or language-specific content space, for instance in the study of Dutch climate change skeptics (in Chapter 3) in the comparison of language versions of a Wikipedia article (as discussed in Chapter 4), and through mention made of places in Twitter hashtag clusters (as described in Chapter 5). However, there are other ways in which content is geo-located (or geo-tagged) on platforms. There are social media platforms that are based centrally on the utility of geo-location, such as Foursquare, a service that allows for 'checking in' on a specific location and thus sharing where you are with your followers, or Instagram, which offers the possibility to give your photo a geo-tag and is for this reason an app often used to share pictures of hotspots in specific places. Here again, social media can offer a lens or window to a specific place, and it could be interesting and productive to ask not only *what kind of place* is this, but also what kinds of mediations of such places, do specific platforms put forward.

As an example outside of my focus on climate change, in a study of the city of Amsterdam through social media data, our work at the Digital Methods initiative recently found that Instagram offers a collected 'boutique view' on the city, while meetup.com (a platform for organizing social gatherings) highlights the 'tech' and 'sports' venues of the city of Amsterdam.[20] For the case study of my book, this means I could select in the Twitter dataset only the geo-demarcated Tweets, or instead look at user-profiles and only select those that state a location. This way, I could research how the state of the debate differs across geo-locations by looking at the origin of a tweet or of the Twitter user (profile). On a methodological level, I could assess the possibilities and limitations of studying place through networked content analysis, assessing how different platforms deal differently with the demarcation of place.[21]

As other technicities add layers to the analysis of an issue or debate, the diversity of content types included in such a 'layered' networked content analysis adds complexity to the analysis. Here, we can learn from controversy mapping, whose scholars have warned against the creation of an all-encompassing 'mother map' that includes all actors, viewpoints, and sources of a certain debate as seen from above.[22] As there is no *above* in controversy mapping, these layers should not be used to create a summary but rather treated as separate

---

20   The layered interactive map is available on: http://bit.ly/amsterdamcartodb, the project page is on the Digital Methods Initiative wiki. Its project page can be found at: https://wiki.digitalmethods.net/Dmi/TheCityAsInterface.
21   A project that explicitly deals with these questions is The Knowledge Mile Atlas, in which I have worked with information designers to create an atlas of a small urban area in Amsterdam. Here, we represent different online data sets of a geographic area by using different methods of geo-demarcation, data analysis, and visualization. First, by geo-locating addresses coming from administrative databases, we showed the density of and the connections between companies registered in the area. Secondly, using natively digital geo-coded objects, such as Foursquare check-ins and geotagged photos, we layered the social media view of the area. Finally, querying street names in the dominant search engine, we collected the online image of each street. Each layer offered a methodological exercise in rethinking geo-location based on the specificity of each platform and the technicity of its content. What is relevant in such methods is the ability to layer the online activity on top of the map of the actual geo-location. The Knowledge Mile maps show the online presence and resonance of an urban area under development in Amsterdam that cuts through the city center and crosses many district and neighborhood 'borders'. Niederer et al. 'Street-Level City Analytics'.
22   Venturini, 'Diving in Magma'.

mappings, in which each offers a detailed window through which we can navigate a debate in all its richness.[23]

In this book, networked content analysis is developed to study the climate debate, a controversy that takes place in science and well beyond, in news media and public debates, and echoing complexly across online platforms. While this book has put forward several research techniques, the example of geo-location indicates that further research will only lead to more material for the content analyst who wants to use networked content for researching debates and controversy. Furthermore, it underlines the need for a more thorough understanding of technicity of content and the adaptive analytical attitude researchers of online networked content need to develop.

The Future of Content: Challenges for Further Research

The biggest challenge for researchers who want to work with networked content may be the multifariousness of content types, data sources, and technicities, which, in order to be compared need to be somehow *comparable*. Here, it is useful to consider how both controversy mapping and digital methods approach this issue. Controversy analysis does not strive for a clean objective picture to arise from the analysis of complex issues and debates. Rather than striving for objectivity, controversy analysis tries to reach what Latour calls 'second-degree objectivity' which is 'the effort to consider as much subjectivity as possible. Unlike first-degree objectivity, which defines a situation of a collective agreement, second-degree objectivity is attained by revealing the full extent of actors' disagreement and is thereby typical of controversial settings.'[24] In second-degree objectivity, it is not necessary to normalize or objectify content in order to make it comparable. Instead, it is the wide array of viewpoints, actors, and sources that build a cartography that Latour himself describes to his students as 'observing and describing'.[25] As controversy mapping does not offer an operationalization of this approach, let alone how to apply it to networked content, it is useful here to look at digital methods for 'cross-platform analysis'.[26]

Digital methods have proposed three approaches to cross-platform analysis, which are strongly related to the methodological difficulties discussed of disentangling content from online platforms. The first approach can be summed up as *medium research* and takes as a point of departure the question of what the platform *does to the content*. How does the platform rank, obfuscate or amplify specific content, and what do we know of its cultures of use? A second approach is that of *social research*. Here, platform technicities are not included in the study, as the researcher focuses on the story told by the content. A third approach is the combination of the two, asking both what the platform does to the content and what stories the content tells.[27] This approach would be most suitable to networked content analysis, where

---

23   Venturini, 'What Is Second-degree Objectivity and How Could It Be Represented'.
24   Venturini, 'Building on Faults', 270.
25   Venturini, 'Building on Faults', 270.
26   Digital Methods Initiative, *DMIR Unit #5*.
27   In networked content analysis, this would be explicitly: 'How does the platform *network* content?'

we could explicitly add how the platforms network content, and how content is 'inter-linked, inter-liked and inter-hashtagged'.[28] However, noting the size of data sets and the necessity of close reading, the scaling up of such methods remains a challenge, which is dealt with by various scholarly fields (ranging from humanities to data science).

The comparability of content from different platforms and the web also becomes an issue in its visualization, or more specifically, in its side-by-side representation in dashboards. As analysts, activists, and decision-makers increasingly make use of dashboards, there is increased urgency to developing critical dashboards, as I alluded to in my mentioning of the *People's Dashboard*. A critical dashboard would show the technicity of content and explain what is left out, what is foregrounded, and what is being amplified by the logic of the platform.

In the preface to his 2010 manifesto *You Are Not a Gadget*, Jaron Lanier writes:

> It is early in the twenty-first century, and that means that these words will mostly be read by nonpersons – automatons or numb mobs composed of people who are no longer acting as individuals. The words will be minced into atomized search-engine keywords within industrial cloud computing facilities located in remote, often secret locations around the world. They will be copied millions of times by algorithms designed to send an advertisement to some person somewhere who happens to resonate with some fragment of what I say.[29]

The future of content presented by Lanier, as material increasingly intertwined with its carriers and platforms, is a future of content networked to the extreme. We will find content made for the network, re-hashed, redistributed, and copied by network infrastructure, and then clicked on, liked, or retweeted by its recipients. The future of content then is content that is written for exponentially *networked technicity*. As content will evolve along with the technicity of its medium, researchers will have to expand our techniques and tools for networked content analysis, continue to develop a critical vocabulary, and produce further concepts and visual languages for the mapping, analysis, and description of networked content.

---

28   Digital Methods Initiative, *DMIR Unit #5*.
29   J. Lanier, *You Are Not a Gadget: A Manifesto,* New York, NY: Alfred A. Knopf, 2010, xiii.

BIBLIOGRAPHY

Adger, W.N. 'Vulnerability', *Global Environmental Change*, 16.3 (2006): 268–281.

Adler, B.T., L. de Alfaro, I. Pye, and V. Raman. 'Measuring Author Contributions to Wikipedia', in *Proceedings of WikiSym 2008, Porto,* New York: ACM, 2008, https://users.soe.ucsc.edu/~luca/papers/08/wikisym08-users.pdf.

Anderegg, W.R., and G.R. Goldsmith. 'Public Interest in Climate Change Over the Past Decade and the Effects of the "Climategate" Media Event', *Environmental Research Letters*, 9.5 (2014): 054005.

Anderson, C. A. 'Heat and violence', *Current Directions in Psychological Science*, 10.1 (2001): 33–38.

Annenberg School for Communication. *George Gerbner Archive*, University of Pennsylvania, 2006.

Annenberg School for Communication. https://www.asc.upenn.edu/.

Auliciems, A., and L. DiBartolo. 'Domestic Violence in a Subtropical Environment: Police Calls and Weather in Brisbane', *International Journal of Biometeorology*, 39.1 (1995): 34–39.

Back, L., C. Lury, and R. Zimmer. 'Doing Real Time Research: Opportunities and Challenges', *National Centre for Research Methods (NRCM)*, *Methodological review paper*, 2012, http://eprints.ncrm.ac.uk/3157/1/real_time_research.pdf.

Baker, N. 'The Charm of Wikipedia,' *New York Review of Books*, 55.4 (2008), http://www.nybooks.com/articles/2008/03/20/the-charms-of-wikipedia/.

Barnett, J. 'Security and Climate Change', *Global Environmental Change*, 13.1 (2003): 7–17.

Barnett, J., and W.N. Adger. 'Climate Change, Human Security and Violent Conflict', *Political Geography*, 26.6 (2008): 639–655.

Barnett, J., S. Lambert, and I. Fry. 'The Hazards of Indicators: Insights from the Environmental Vulnerability Index', *Annals of the Association of American Geographers*, 98.1 (2008): 102–119.

Beck, U. *World at Risk*, Cambridge: Polity Press, 2009.

Beer, D. 'Power Through the Algorithm? Participatory Web Cultures and the Technological Unconscious', *New Media & Society*, 11.6 (2009): 985–1002.

Benkler, Y. *The Wealth of Networks: How Social Production Transforms Markets and Freedom*, New Haven: Yale University Press, 2006.

Berelson, B. 'Content Analysis in Communication Research', 1952, http://psycnet.apa.org/psycinfo/1953-07730-000

Berelson, B. and P.J. Salter. 'Majority and Minority Americans: An Analysis of Magazine Fiction', *The Public Opinion Quarterly*, 10 (1948): 168–190.

Bilic, P., and L. Bulian. 'Lost in Translation: Contexts, Computing, Disputing on Wikipedia,' in *iConference 2014*, 2014, pp. 32–44.

Birkmann, J. 'Measuring Vulnerability to Promote Disaster-resilient Societies: Conceptual Frameworks and Definitions,' *Measuring Vulnerability to Natural Hazards: Towards Disaster Resilient Societies*, 1, (2006): 9–54.

Blakeslee, D., and R. Fishman. 'Rainfall shocks and property crimes in agrarian societies: Evidence from India', 2013, http://papers.ssrn.com/sol3/papers.cfm?abstract_id=2208292.

Blondel, V.D., J.-L. Guillaume, R. Lambiotte, and E. Lefebvre. 'Fast Unfolding of Communities in Large Networks', *Journal of Statistical Mechanics: Theory and Experiment* (2008).

Bohlken, A. T., and E.J. Sergenti. 'Economic Growth and Ethnic Violence: An Empirical Investigation of Hindu-Muslim Riots in India', *Journal of Peace Research*,47 (2010): 589–600.

Borra, E., and B. Rieder. 'Programmed Method: Developing a Toolset for Capturing and Analyzing Tweets', *Aslib Journal of Information Management* 66.3 (2014): 262–278.

Borra, E., E. Weltevrede, P. Cuiccarelli, A. Kaltenbrunner, D. Laniado, G. Magni, and T. Venturini. 'Societal Controversies in Wikipedia Articles', in *CHI'15: 33rd Annual ACM Conference on Human Factors in Computing Systems*, 2015, http://www.contropedia.net/publications/Borra%20et%20al.%20-%202015%20-%20Societal%20Controversies%20in%20Wikipedia%20Articles%20-%20post-print.pdf.

boyd, danah, S. Golder, and G. Lotan. 'Tweet, Tweet, Retweet: Conversational Aspects of Retweeting on Twitter', in *43rd Hawaii International Conference on System Sciences (HICSS)*, 2010.

Brin, S. and L. Page, L. 'The Anatomy of a Large-Scale Hypertextual Web Search Engine', *Computer Networks*, 56.18 (2010): 3825–3833.

Brossard, D., J. Shanahan, and K. McComas. 'Are Issue-cycles Culturally Constructed? A Comparison of French and American Coverage of Global Climate Change', *Mass Communication & Society*, 7.3 (2004): 359–377.

Brugh, M. aan de. 'Liberaal in de Strijd Tegen Klimaatgekte', *NRC Handelsblad*, 19 February 2007, http://vorige.nrc.nl/article1771418.ece.

Bruns, A. and Y.E. Liang, 'Tools and Methods for Capturing Twitter Data During Natural Disasters', *First Monday* 17.4 (2012): http://firstmonday.org/ojs/index.php/fm/article/view/3937/3193.

Bruns, A. G*atewatching: Collaborative Online News Production*, New York: Peter Lang, 2005.

Bruns, A. 'Methodologies for Mapping the Political Blogosphere: Explorations Using the Issue-Crawler Research Tool', *First Monday*, 12.5 (2007): http://firstmonday.org/ojs/index.php/fm/article/view/1834/

Bruns, A. *Blogs, Wikipedia, Second Life, and Beyond: From Production to Produsage*, New York: Peter Lang, 2008.

Bruns, A. and J.E. Burgess. 'The use of Twitter Hashtags in the Formation of Ad Hoc Publics', in *Proceedings of the 6th European Consortium for Political Research (ECPR) General Conference 2011*, 2011, http://eprints.qut.edu.au/46515.

Burke, M. and R. Kraut, R. 'Taking Up the Mop: Identifying Future Wikipedia Administrators' in *Proceedings of the 2008 CHI Conference, Florence*, New York: ACM, 2008, pp. 3441-3446, http://portal.acm.org/citation.cfm?id=1358628.1358871.

Burkhart, K. 'Organizing for Action Delivers Unicorn Trophies to 135 Climate Deniers in Congress', *The Huffington Post*, 13 August 2013, http://www.huffingtonpost.com/2013/08/13/organizing-for-action-climate-deniers_n_3750126.html.

Cambrosio, A., P. Cottereau, S. Popowycz, A. Mogoutov, and T. Vichnevskaia. 'Analysis of Heterogenous Networks: The ReseauLu Project', in B. Reber and C. Brossaud (eds) *Digital Cognitive Technologies: Epistemology and the Knowledge Economy*, Hoboken, NJ: John Wiley & Sons, Inc, 2013.

Card, S.K., J.D. Mackinlay, and B. Shneiderman. *Readings in Information Visualization: Using Vision to Think*, San Francisco, CA: Morgan Kaufmann, 1999.

Carr, N. *The Big Switch: Rewiring the World, from Edison to Google*, New York, NY: W.W. Norton & Company, 2008.

Carvalho, A. and J. Burgess. 'Cultural Circuits of Climate Change in U.K. Broadsheet Newspapers, 1985-2003', *Risk Analysis*, 25.6 (2005): 1457–1469.

Cattaneo, C. 'As B.C. Election Looms, Both NDP and Liberals Take Hard Line on Oil Pipelines', *Financial Post*, 2013, http://business.financialpost.com/2013/05/06/bc-election-oil-sands/?__lsa=d52e-3289.

Citizendium. 'Citizendium Beta', http://en.citizendium.org/wiki/Welcome_to_Citizendium.

'Content Segmentation: Differentiate Your Brand Online', 5 April 2012, http://contentmarket-inginstitute.com/2012/04/use-content-segmentation-to-differentiate-your-brand/.

Cook, J., D. Nuccitelli, S.A. Green, M. Richardson, B. Winkler, R. Painting, and A. Skuce. 'Quantifying the Consensus on Anthropogenic Global Warming in the Scientific Literature', *Environmental Research Letters*, 8.2 (2013): 024024.

Dalby, A. *The World and Wikipedia: How We Are Editing Reality*. Somerset: Siduri Books, 2009.

Deibert, R., J. Palfrey, R. Rohozinski, J. Zittrain, and M. Haraszti. *Access Controlled: The Shaping of Power, Rights, and Rule in Cyberspace*, Cambridge, MA: MIT Press, 2010.

Deibert, R., J. Palfrey, R. Rohozinski, J. Zittrain, and J.G. Stein. *Access Denied: The Practice and Policy of Global Internet Filtering*, Cambridge, MA: MIT Press, 2008.

Delbecq, D. 'A [F]rench Climate Skeptic Comes Out: He Is a Physicist', *Effets de Terre*, 2010, http://effetsdeterre.fr/2010/04/21/a-french-climate-skeptic-comes-out-he-is-a-physicist/.

Delbecq, D. 'Dossier Climato-sceptiques', *TerraEco* (April 2010): 50–62.

Delbecq, D. and S. Niederer. 'Climatosceptiques et Climatologues, Quelle Place sur l'Inter-net?', 2010, http://effetsdeterre.fr/2010/04/12/climatosceptiques-quelle-place-sur-linternet/.

Deleuze, G. 'Society of Control', *L'autre Journal,* 1 (1990): http://www.nadir.org/nadir/archiv/netzkritik/societyofcontrol.html

Denning, P., J. Horning, D. Parnas, and L. Weinstein. 'Wikipedia Risks', *Communications of the ACM*, 48.12 (2005): 152.

Digital Methods Initiative. 'Issue Image Analysis', 2007, https://wiki.digitalmethods.net/Dmi/IssueImageAnalysis.

Digital Methods Initiative. 'Issue Animals Research', 2008, https://www.digitalmethods.net/Dmi/IssueImageAnalysis.

Digital Methods Initiative. 'Networked Content', 2008, https://digitalmethods.net/Digitalmeth-ods/TheNetworkedContent.

Digital Methods Initiative. 'The Place of Issues', 2009, https://wiki.digitalmethods.net/Dmi/ThePlaceOfIssues.

Digital Methods Initiative. 'The City as Interface', 2014, https://wiki.digitalmethods.net/Dmi/TheCityAsInterface.

Digital Methods Initiative. *DMIR Unit #5: Cross-Platform Analysis*, Amsterdam: University of Amsterdam, 2015.

Digital Methods Initiative. 'The People's Dashboard', 2015, https://wiki.digitalmethods.net/Dmi/PeoplesDashboard.

Digital Methods Initiative. 'Climate Change Skeptics in the Wikipedia Climate Change Space', https://wiki.digitalmethods.net/Dmi/WikipediaClimateChangeSkeptics.

Digital Methods Initiative. 'Lippmannian Device', https://wiki.digitalmethods.net/Dmi/ToolLippmannianDevice.

Djerf-Pierre, M. 'When Attention Drives Attention: Issue Dynamics in Environmental News Reporting Over Five Decades', *European Journal of Communication*, 27.3 (2012): 291–304.

Doucleff, M. 'Could Hotter Temperatures From Climate Change Boost Violence?', 2 Augusts 2013, http://www.npr.org/blogs/health/2013/08/02/208218317/could-hotter-temperatures-from-climate-change-boost-violence.

Downs, A. 'Up and Down with Ecology: The Issue-attention Cycle', *The Public Interest* 28 (1972): 38.

EMAPS. 'Vulnerability, Resilience and Conflict: Mapping Climate Change, Reading Cli-fi', *Electronic Maps to Assist Public Science Blog*, 2013, http://www.emapsproject.com/blog/archives/2293.

EMAPS. 'Climaps: A Global Issue Atlas of Climate Change Adaptation', 2014, http://climaps.eu/.

EMAPS. 'Climate Change Tweets Co-Hashtag Cluster Map', 2014, http://climaps.eu/#!/map/profiling-adaptation-and-its-place-in-climate-change-debates-with-twitter-ii.

EMAPS. 'Profiling Adaptation And Its Place In Climate Change Debates With Twitter', 2014, http://climaps.eu/#!/map/profiling-adaptation-and-its-place-in-climate-change-debates-with-twitter-I.

EMAPS. 'Reading the State of Climate Change From Digital Media', 2014, http://climaps.eu/#!/narrative/reading-the-state-of-climate-change-from-digital-media.

EMAPS. 'Reading the State of Climate Change From the Web: Top Google Results', 2014, http://climaps.eu/#!/map/profiling-adaptation-and-its-place-in-climate-change-debates-with-twitter-ii.

EMAPS. 'Who Deserved to Be Funded? A Closer Look at the Practices of Vulnerability Assessment and the Priorities of Adaptation Funding', 2014, http://climaps.eu/#!/narrative/who-deserves-to-be-funded.

EMAPS. 'Who is Vulnerable According to Whom?' 2014, http://climaps.eu/#!/map/who-is-vulnerable-according-to-whom.

EMAPS. 'Contropedia', 2015, http://contropedia.net/.

Eriksen, S.H. and P.M. Kelly, P. M. 'Developing Credible Vulnerability Indicators for Climate Adaptation Policy Assessment', *Mitigation and Adaptation Strategies for Global Change* 12.4 (2007): 495–524.

Galloway, A. *Protocol: How Control Exists after Decentralization*, Cambridge, MA: MIT Press, 2004.

Gayo-Avello, D. 'I Wanted to Predict Elections with Twitter and All I Got Was This Lousy Paper: A Balanced Survey on Election Prediction Using Twitter Data', *Arxiv Preprint arXiv12046441*, 2012, http://arxiv.org/pdf/1204.6441.pdf.

Geiger, R.S. and D. Ribes. 'The Work of Sustaining Order in Wikipedia: The Banning of A Vandal', in *Proceedings of the ACM 2010 conference on Computer supported cooperative work (CSCW)*, Atlanta, GA: Association for Computing Machinery, 2010, http://www.stuartgeiger.com/wordpress/wp-content/uploads/2009/10/cscw-sustaining-order-wikipedia.pdf.

Gerbner, G. 'Toward "Cultural Indicators": The Analysis of Mass Mediated Public Message Systems', *Educational Technology Research and Development* 17.2 (1969): 137–148.

Gerbner, G. 'Cultural Indicators: The Case of Violence in Television Drama', *The Annals of the American Academy of Political and Social Science* 388.1 (1970): 69–81.

Gerbner, G., L. Gross, N. Signorielli, M. Morgan, and M. Jackson-Beeck. 'The Demonstration of Power: Violence Profile No. 10', *Journal of Communication* 29.3 (1979): 177–196.

Gerbner, G., O. Holsti, K. Krippendorff, W.J. Paisley, and P.J. Stone (eds) *The Analysis of Communication Contents: Development in Scientific Theories and Computer Techniques*, Wiley, 1969, http://www.sidalc.net/cgi-bin/wxis.exe/?IsisScript=UACHBC.xis&method=post&formato=2&cantidad=1&expresion=mfn=017033.

Gerlitz, C. and A. Helmond. 'The Like Economy: Social Buttons and the Data-intensive Web', *New Media & Society*, 2013, http://nms.sagepub.com/content/early/2013/02/03/1461444812472322

Gerlitz, C. and B. Rieder. 'Mining One Percent of Twitter: Collections, Baselines, Sampling', *M/C Journal*, 16.2 (2013).

Ghosh, R.A. and V.V. Prakash. 'Orbiten Free Software Survey', *First Monday* 5.7 (2000): http://www.firstmonday.org/issues/issue5_7/ghosh/

Gillis, J. and C. Krauss. 'Exxon Mobil Investigated for Possible Climate Change Lies by New York Attorney General', *The New York Times*, 5 November 2015, http://www.nytimes.com/2015/11/06/science/exxon-mobil-under-investigation-in-new-york-over-climate-statements.html.

Gladwell, M. 'Small Change: Why the Revolution Will Not Be Tweeted', *New Yorker*, 4 October 2010, http://www.newyorker.com/reporting/2010/10/04/101004fa_fact_gladwell

Govcom.org Foundation. 'IssueCrawler: Instructions of Use', http://www.govcom.org/Issuecrawler_instructions.html.

Greenpeace, http://www.exxonsecrets.org/maps.php.

Griffith, V. 'Wikiscanner Homepage', 2007-2008, http://virgil.gr/31.html.

Halfaker, A., R.S. Geiger, J. Morgan, and J. Riedl. 'The Rise and Decline of an Open Collaboration System: How Wikipedia's Reaction to Sudden Popularity Is Causing Its Decline', *American Behavioral Scientist* 57.5 (2013): 664–688, http://doi.org/10.1177/0002764212469365.

Helmond, A. 'The Perceived Freshness Fetish', 2007, http://www.annehelmond.nl/wordpress/wp-content/uploads/2007/06/annehelmond_pff.pdf.

Helmond, A. *The Web as Platform: Data Flows in Social Media,* Ph.D. Thesis, 19 June 2015, University of Amsterdam, Amsterdam.

Hendrix, C.S. and I. Salehyan. 'Climate Change, Rainfall, and Social Conflict in Africa', *Journal of Peace Research* 49.1 (2012): 35–50.

Hermida, A. 'Twittering the News', *Journalism Practice* 4.3 (2010): 297–308.

Herring, S. 'Web Content Analysis: Expanding the Paradigm', in J. Hunsinger et al. (eds) *International Handbook of Internet Research*, Dordrecht: Springer, 2010, pp. 233-249.

Hinkel, J. 'Indicators of Vulnerability and Adaptive Capacity: Towards a Clarification of the Science–policy Interface', *Global Environmental Change* 21.1 (2011): 198–208.

Hoffman, A. J. 'Talking Past Each Other? Cultural Framing of Skeptical and Convinced Logics in the Climate Change debate', *Organization Environment* 24.1 (2011): 3–33.

Hoggan, J. and R. Littlemore. *Climate Cover-up: The Crusade to Deny Global Warming*, Vancouver: Greystone Books, 2009.

Homer-Dixon, T. F. 'On the Threshold: Environmental Changes as Causes of Acute Conflict', *International Security* 16 (1991): 76–116.

Howard, J. 'Climate Change Mitigation and Adaptation in Developed Nations: A Critical Perspective on the Adaptation Turn in Urban Climate Planning', in S. Davoudi, J. Crawford and A. Mehmood (eds) *Planning for Climate Change: Strategies for Mitigation and Adaptation for Spatial Planning*, London: Earthscan, 2009, pp. 19-32.

Howe, J. 'The Rise of Crowdsourcing', *Wired Magazine* 14.6 (2006): 1–4.

Hsiang, S.M., M. Burke, E. Miguel. 'Quantifying the Influence of Climate on Human Conflict', *Science*, 341.6151 (2013): http://doi.org/10.1126/science.1235367.

IPCC. 'Working Group II: Impacts, Adaptation and Vulnerability: Summary for Policymakers', 2001, http://www.ipcc.ch/ipccreports/tar/wg2/index.php?idp=8.

IPCC. 'Climate Change 2014: Synthesis Report: Contribution of Working Groups I, II and III to the Fifth Assessment Report of the Intergovernmental Panel on Climate Change', Geneva: IPCC, 2014, http://ar5-syr.ipcc.ch/.

Jemielniak, D. *Common Knowledge? An Ethnography of Wikipedia*, Stanford, CA: Stanford University Press, 2014.

Jenkins, H. *Convergence Culture: Where Old and New Media Collide.* Cambridge, MA: MIT Press, 2006.

Keen, A. *The Cult of the Amateur: How blogs, MySpace, YouTube, and the Rest of Today's User-generated Media Are De-stroying Our Economy, Our Culture, and Our Values*, New York: Doubleday Currency, 2007.

Kenrick, D.T. and S.W. MacFarlane, S. W. 'Ambient Temperature and Horn Honking: A Field Study of the Heat/Aggression Relationship', *Environment and Behavior* 18 (1986): 179–197.

Kittur, A., E. Chi, B.A. Pendleton, B. Suh, and T. Mytkowicz. 'Power of the Few vs. Wisdom of the Crowd: Wikipedia and the Rise of the Bourgeoisie', in *CHI*, 2007, San Jose.

Kittur, A., and R.E. Kraut. 'Harnessing the Wisdom of Crowds in Wikipedia: Quality Through Coordination', in *Proceedings of the ACM 2008 Conference on Computer Sup-ported Cooperative Work*, New York: ACM, 2008, pp. 37-46.

Klein, N. *This Changes Everything*, New York: Simon and Schuster, 2014.

Klein, R. J. T. 'Identifying Countries That Are Particularly Vulnerable to the Adverse Effects of Climate Change: An Academic or a Political Challenge?' *Carbon and Climate Law Review* 3 (2009): 284–291.

KNAW. *Klimaatverandering, Wetenschap en Debat,* Amsterdam: Koninklijke Nederlandse Academie van Wetenschappen, 2011, https://www.knaw.nl/nl/actueel/publicaties/klimaat-verandering-wetenschap-en-debat/@@download/pdf_file/20101047.pdf.

Kohut, A., D.C. Doherty, M. Dimock, M., and S. Keeter, S. 'Fewer Americans See Solid Evidence of Global Warming', *Washington, DC: Pew Research Center,* 2009.

König, R., M. Rasch (eds) *Society of the Query Reader: Reflections on Web Search,* Amsterdam: Institute of Network Cultures, 2014.

Krippendorff, K. *Content Analysis: An Introduction to its Methodology,* first edition, Beverly Hills, CA: Sage Publications, 1980.

Krippendorff, K. *Content Analysis: An Introduction to its Methodology,* second edition, Thousand Oaks: Sage Publications, 2004.

Krippendorff, K. *Content Analysis: An Introduction to its Methodology,* third edition, Thousand Oaks, CA: Sage Publications, 2013.

Labohm, H. 'Klimaatsceptici Verzoeken KNAW Klimaatrapport in te Trekken', October 2011, http://www.dagelijksestandaard.nl/2011/10/klimaatsceptici-verzoeken-knaw-klimaatrap-port-in-te-trekken.

Labohm, H., S. Rozendaal, and D. Thoenes. *Man-Made Global Warming: Unravelling a Dogma,* Essex: Multi-Science Publishing Co. Ltd, 2004.

Lanier, J. *You Are Not a Gadget: A Manifesto,* New York, NY: Alfred A. Knopf, 2010.

Latour, B. 'Technology Is Society Made Durable', in J. Law (ed.) *A Sociology of Monsters Essays on Power, Technology and Domination,* London: Routledge, 1991, pp. 103-132.

Latour, B. *Reassembling the Social: An Introduction to Actor-Network-Theory,* Oxford: Oxford University Press, 2005.

Latour, B. 'Mapping Controversies', presented at the Digital Methods Summer School, University of Amsterdam, Amsterdam, 2007.

Latour, Bruno. *Reassembling the Social,* Oxford: Oxford University Press, 2005.

Latour, B. 'Waiting for Gaia: Composing the Common World Through Arts and Politics', *Equilibri* 16.3 (2012): 515–538.

Lazer, D., R. Kennedy, G. King, and A. Vespignani. 'The Parable of Google Flu: Traps in Big Data Analysis', *Science* 343 (2014): 1203–1205.

Leskovec, J., L. Backstrom, and J. Kleinberg. 'Meme-tracking and the Dynamics of the News Cycle', In *Proceedings of the 15th ACM SIGKDD international conference on Knowledge discovery and data mining*, ACM, 2009, pp. 497-506, http://dl.acm.org/citation.cfm?id=1557077.

Lih, A. *The Wikipedia Revolution: How a Bunch of Nobodies Created the World's Greatest Encyclopedia*, London: Aurum Press, 2009.

Lotan, G., E. Graeff, M. Ananny, D. Gaffney, I. Pearce, I. and danah boyd. 'The Revolutions Were Tweeted: Information Flows During the 2011 Tunisian and Egyptian Revolutions', *International Journal of Communication* 5 (2011): 1375–1405.

Lovink, G. 'The Society of the Query and the Googlisation of Our Lives: A Tribute to Joseph Weizenbaum', *Eurozine*, 2008, http://www.eurozine.com/articles/2008-09-05-lovink-en.html

Lovink, G. *Social Media Abyss: Critical Internet Cultures and the Force of Negation*, Cambridge, UK: Polity Press, 2016.

MacNab, J. 'Will climate be a winner in British Columbia's election?' *Pembina Institute*, 2013, http://www.pembina.org/blog/724.

Mann, M.E., R.S. Bradley, and M.K. Hughes, M. K. 'Northern Hemisphere Temperatures During the Past Millennium: Inferences, Uncertainties, and Limitations', *Geophysical Research Letters* 26.6 (1999): 759–762.

Marres, N. 'There is Drama in Networks', in J. Brouwer and A. Mulder (eds) *Interact or Die!*, Rotterdam: NAI Publishers, 2007, pp. 174-187.

Marres, N. 'Why Map Issues? On Controversy Analysis as a Digital Method', *Science, Technology & Human Values*, 0162243915574602, 2015, http://doi.org/10.1177/0162243915574602.

Marres, N. and D. Moats. 'Mapping Controversies with Social Media: The Case for Symmetry', *Social Media + Society* 1.2 (2015): 2056305115604176, http://doi.org/10.1177/2056305115604176.

Marres, N. and E. Weltevrede. 'Scraping the Social? Issues in Live Social research', *Journal of Cultural Economy* 6.3 (2013): 313–335.

Martin, Shawn W., Michael Brown, Richard Klavans, and Kevin Boyak. 'OpenOrd: An Open-source Toolbox for Large Graph Layout', *Proceedings of the SPIE Visualisation and Data Analysis*, 2011, https://doi.org/10.1117/12.871402.

McComas, K. and J. Shanahan. 'Telling Stories About Global Climate Change Measuring the Impact of Narratives on Issue Cycles', *Communication Research* 26.1 (1999): 30–57.

McCright, A. M. and R.E. Dunlap. 'Defeating Kyoto: The Conservative Movement's Impact on US Climate Change Policy', *Social Problems* 50.3 (2003): 348–373.

McMillan, S. 'The Microscope and the Moving Target: The Challenge of Applying Content Analysis to the World Wide Web', *Journalism and Mass Communication Quarterly* 77 (2000): 80–88.

Mehlum, H., E. Miguel, and R. Torvik. 'Poverty and Crime in 19th Century Germany', *Journal of Urban Economics* 59.3 (2006): 370–388.

Merriam-Webster, A. *Webster's New Collegiate Dictionary.* G.&C. Merriam Company, Publishers, 1961.

Michaels, D. *Doubt is Their Product: How Industry's Assault on Science Threatens Your Health*, Oxford: Oxford University Press, 2008, https://books.google.nl/books?hl=nl&lr=&id=Qedeu_5_IEEC&oi=fnd&pg=PT8&dq=doubt+is+their+product&ots=d9HaBSwLOn&sig=a9vUIiKcNv-vHDZ3pKzhD0GkmCd4.

Moats, D. 'From Digital Methods to Digital Ontologies: Bruno Latour and Richard Rogers at CSISP', 2012, http://www.csisponline.net/2012/03/12/from-digital-methods-to-digital-ontologies-bruno-latour-and-richard-rogers-at-csisp/.

Morozov, E. 'Iran: Downside to the "Twitter Revolution",' *Dissent* 56.4 (2009): 10–14.

Moser, S. C. 'Costly Knowledge – Unaffordable Denial: The Politics of Public Understanding and Engagement on Climate Change', in *The Politics of Climate Change: A Survey*, 2010, 155–181.

Mother Jones. 'Put a Tiger in Your Think Tank.' *Mother Jones,* 2005, http://www.motherjones.com/politics/2005/05/put-tiger-your-think-tank.

Nacu-Schmidt, A., K. Andrews, M. Boykoff, M. Daly, L. Gifford, G. Luedecke, and L. McAllister. 'World Newspaper Coverage of Climate Change or Global Warming, 2004-2016', 2016, http://sciencepolicy.colorado.edu/media_coverage.

Nerlich, B. '"Climategate": Paradoxical Metaphors and Political Paralysis', *Environmental Values* 19.4 (2010): 419–442.

Niederer, S. 'Climate Change Skeptics in Science', 2009, http://www.mappingcontroversies.net/Home/PlatformClimateChangeSkepticsScience.

Niederer, S. 'Global Warming Is Not a Crisis! Studying Climate Change Skepticism on the Web', *Necsus* 3 (Spring 2013): http://www.necsus-ejms.org/global-warming-is-not-a-crisis-studying-climate-change-skepticism-on-the-web/.

Niederer, S. 'Interview', in N. Nova and F. Kaplan (eds) *Wikipedia's Miracle*, Lausanne: EPFL Press, 2016, pp. 53-61.

Niederer, S., G. Colombo, M. Mauri, and M. Azzi, M. 'Street-Level City Analytics: Mapping the Amsterdam Knowledge Mile' in *Hybrid City 2015: Data to the People*, Athens: University of Athens, 2015, www.media.uoa.gr/hybridcity.

Niederer, S., G. Colombo, M. Mauri, and M. Azzi, M. 'Street-Level City Analytics: Mapping the Amsterdam Knowledge Mile', in *Hybrid City 2015: Data to the People*, Athens: University of Athens, 2015, www.media.uoa.gr/hybridcity.

Niederer, S. and R. Taudin Chabot, R. 'Deconstructing the Cloud: Responses to Big Data Phenomena From Social Sciences, Humanities and the Arts', *Big Data & Society* 2.2 (2015): http://doi.org/10.1177/2053951715594635.

Niederer, S. and J. van Dijck. 'Wisdom of the Crowd or Technicity of Content? Wikipedia as a Sociotechnical System', *New Media & Society* 12.8 (2010): 1368–1387.

Niederer, S. and J. van Dijck, J. 'Wisdom of the Crowd or Technicity of Content? Wikipedia as a Sociotechnical System', in M. David and P. Milward (eds) *Researching Society Online*, London: Sage, 2014.

Nieuwspoort, http://www.nieuwspoort.nl/over-nieuwspoort/.

Nisbet, M.C. and T. Myers, T. 'The Polls-Trends: Twenty Years of Public Opinion About Global Warming', *Public Opinion Quarterly* 71.3 (2007): 444–470.

NPR. 'How Could a Drought Spark a Civil War?', 2013, http://www.npr.org/2013/09/08/220438728/how-could-a-drought-spark-a-civil-war.

O'Neill, S.J., M. Boykoff, S. Niemeyer, and S.A. Day. 'On the Use of Imagery for Climate Change Engagement', *Global Environmental Change* 23.2 (2013): 413–421.

O'Neil, M. *Cyberchiefs: Autonomy and Authority in Online Tribes*, New York: Pluto Press, 2009.

Oreskes, N. 'The Scientific Consensus on Climate Change', *Science* 306.5702 (2004): 1686–1686.

Oreskes, N. 'Beyond the Ivory Tower: The Scientific Consensus on Climate Change', *Science* 206.5702 (2007): 1686.

Oreskes, N. 'The Scientific Consensus on Climate Change: How Do We Know We're Not Wrong?' in J.F.C. DiMento and P. Doughman (eds) *Climate Change: What It Means for Us, Our Children, and Our Grandchildren*, Cambridge: MIT Press, 2007, pp. 65-99.

Page, Lawrence, Sergey Brin, Rajeev Motwani, and Terry Winograd. 'The PageRank Citation Ranking: Bringing Order to the Web.' Technical Report, Stanford InfoLab, 1999, http://ilpubs. stanford.edu:8090/422/.

Pearce, F. 'The Five Key Leaked Emails From UEA's Climatic Research Unit', *The Guardian*, 7 July 2010, http://www.theguardian.com/environment/2010/jul/07/hacked-climate-emails-analysis.

Perez, I. 'Climate Change and Rising Food Prices Heightened Arab Spring', 2013, http://www. scientificamerican.com/article/climate-change-and-rising-food-prices-heightened-arab-spring/.

Pfeil, U., P. Zaphiris, C.S. Ang. 'Cultural Differences in Collaborative Authoring of Wikipedia', *Journal of Computer-Mediated Communication* 12.1 (2006): 88–113.

Phelan, O., K. McCarthy, and B. Smyth. 'Using Twitter to Recommend Real-time Topical News', in *Proceedings of the Third ACM Conference on Recommender Systems*, 2009, pp. 385-388, http://dl.acm.org/citation.cfm?doid=1639714.1639794.

Pinch, T. and C. Leuenberger. 'Studying Scientific Controversy from the STS Perspective: Concluding Remarks on Panel "Citizen Participation and Science and Technology"', in *East Asian Science, Technology and Society*, 2006, http://fr.curriculumforge.org/TravaillongVincen-tr?action=AttachFile&do=get&target=Pinch+studying.pdf

Platform Politics. '*Platform Politics: Call for Papers: A Multidisciplinary Conference*', Cambridge, UK, 2011, http://www.networkpolitics.org/content/platform-politics-call-papers.

Plumer, B. 'Drought Helped Cause Syria's War: Will Climate Change Bring More Like It?', *The Washington Post*, 10 September 2013, https://www.washingtonpost.com/news/wonk/wp/2013/09/10/drought-helped-caused-syrias-war-will-climate-change-bring-more-like-it/.

Poell, T. and K. Darmoni. 'Twitter as a Multilingual Space: The Articulation of the Tunisian Revolution Through #sidibouzid', *NECSUS European Journal of Media Studies* 1.1 (2012): 14–34.

Poe, M. 'The Hive', *The Atlantic Online*, September 2006, http://www.theatlantic.com/doc/200609/wikipedia.

Raleigh, C. and H. Urdal, H. 'Climate Change, Environmental Degradation and Armed Conflict', *Political Geography* 26.6 (2007): 674–694.

Reagle, J.M. *Good Faith Collaboration: The Culture of Wikipedia*, Cambridge, MA: MIT Press, 2010.

Renn, O. and P. Graham. *White Paper on Risk Governance: Towards an Integrative Approach*, International risk governance council, 2015.

Reuveny, R. 'Climate Change-induced Migration and Violent Conflict', *Political Geography* 26.6 (2007): 656–673.

Rieder, B. 'What is in PageRank? A Historical and Conceptual Investigation of a Recursive Status Index', *Computational Culture*, 2012, http://doi.org/http://computationalculture.net/article/what_is_in_pagerank.

Rifkin, Jeremy. *The Hydrogen Economy: The Creation of the Worldwide Energy Web and the Redistribution of Power on Earth Tarcher.* New York: Putnam, 2002.

Rogers, R. *Information Politics on the Web*, Cambridge, MA: MIT Press, 2004.

Rogers, R. *The End of the Virtual: Digital Methods*, Amsterdam: Vossiuspers UvA, 2009.

Rogers, R. *Digital Methods*, Cambridge, MA.: MIT Press, 2013.

Rogers, R. 'Debanalising Twitter: The Transformation of an Object of Study', in K. Weller, A. Bruns, J. Burgess, M. Mahrt, and C. Puschmann (eds) *Twitter & Society*, New York: Peter Lang, 2014, pp. xi-xxvi.

Rogers, R., F. Janssen, M. Stevenson, and E. Weltevrede. 'Mapping Democracy', in *Global Informaton Society Watch*, The Hague: Hivos, 2009, pp. 47-57.

Rogers, R. and N. Marres, N. 'Landscaping Climate Change: A Mapping Technique for Understanding Science and Technology Debates on the World Wide Web', *Public Understanding of Science* 9.2 (2000): 141–163.

Rogers, R. and E. Sendijarevic. 'Neutral or National Point of View? A Comparison of Screbrenica Articles across Wikipedia's Language Versions', presented at the Wikipedia Academy 2012, Berlin, 2012.

Rogers, R., E. Weltevrede, S. Niederer, and E. Borra. 'National Web Studies: The case of Iran', in J. Hartley, J. Burgess and A. Bruns (eds) *Blackwell Companion to New Media Dynamics*, Oxford: Blackwell, 2013, pp. 142-166

Rosenzweig, R. 'Can History Be Open Source? Wikipedia and the Future of the Past', *The Journal of American History* 93.1 (2006): 117–146.

'Saskatchewan Citizens' Hearings on Climate Change', 2014, http://skclimatehearings.org/.

Savage, M. 'Contemporary Sociology and the Challenge of Descriptive Assemblage', *European Journal of Social Theory* 12.1 (2009): 155–174, http://doi.org/10.1177/1368431008099650.

Schmidt, C. W. 'A Closer Look at Climate Change Skepticism', *Environmental Health Perspectives* 118.12 (2010): A536–A540.

Scrinzi, F. and P. Massa, P. 'WikiWatchDog', 2010, http://www.wikiwatchdog.com.

Shirky, C. *Here Comes Everybody: The Power of Organizing without Organizations*, New York: Penguin Press, 2008.

Shirky, C. 'A Speculative Post on the Idea of Algorithmic Authority', 2009, http://www.shirky.com/weblog/2009/11/a-speculative-post-on-the-idea-of-algorithmic-authority/.

Simonite, T. 'The Decline of Wikipedia', *MIT Technology Review*, 2013, http://www.technologyreview.com/featuredstory/520446/the-decline-of-wikipedia/.

Skeptical Science. 'The Consensus Project', 2013, http://theconsensusproject.com/.

Smyrnaios, N. and B. Rieder. 'Social Infomediation of News on Twitter: A French Case Study', *NECSUS* (Autumn 2013), http://www.necsus-ejms.org/social-infomediation-of-news-on-twitter-a-french-case-study/.

Sprenger, T.O., A. Tumasjan, P.G. Sandner, and I.M. Welpe. 'Tweets and Trades: the Information Content of Stock Microblogs', *European Financial Management* 20 (2014): 926-957, 10.1111/j.1468-036X.2013.12007.x.

Stalder, F. and J. Hirsh. 'Open source intelligence', *First Monday* 7.6 (2002): http://firstmonday.org/htbin/cgiwrap/bin/ojs/index.php/fm/article/viewArticle/961/882.

Stanton, E.A., J. Cegan, R. Bueno, and F. Ackerman. *Estimating Regions' Relative Vulnerability to Climate Damages in the CRED Model*, Somerville, MA: Stockholm Environment Institute, 2011, http://www.sei-international.org/mediamanager/documents/Publications/Climate-mitigation-adaptation/Economics_of_climate_policy/sei-workingpaperus-1103.pdf.

Stevenson, R.W. and J.M. Broder. 'Climate Change Prominent in Obama's Inaugural Address', *The New York Times*, 21 January 2013, http://www.nytimes.com/2013/01/22/us/politics/climate-change-prominent-in-obamas-inaugural-address.html.

Sullivan, A. 'The Revolution Will Be Twittered', *The Atlantic*, 2009, http://www.theatlantic.com/daily-dish/archive/2009/06/the-revolution-will-be-twittered/200478/.

Sunstein, C.R. *Infotopia: How Many Minds Produce Knowledge*, Oxford: Oxford University Press, 2006.

Surowiecki, J. *The Wisdom of Crowds: Why the Many Are Smarter than the Few and How Collective Wisdom Shapes Business, Societies and Nations,* New York: Doubleday, 2004.

Swartz, A. 'Who Writes Wikipedia', 2006, http://www.aaronsw.com/weblog/whowriteswikipedia/.

Tapscott, D. and A.D. Williams. *Wikinomics. How Mass Collaboration Changes Everything*, New York: Penguin, 2006.

Taylor, L. 'Rudd Accuses Abbott of Abandoning Australia's Alimate Commitments, *The Guardian*, 13 September 2013, http://www.theguardian.com/world/2013/sep/03/rudd-abbott-abandoning-climate-commitments.

The Heartland Institute. 'First International Conference on Climate Change (ICCC-1)', 2008, http://climateconferences.heartland.org/iccc1/.

Tkacz, N. *Wikipedia and the Politics of Openness*, Chicago: University of Chicago Press, 2015.

Tufekci, Z. and C. Wilson, C. 'Social Media and the Decision to Participate in Political Protest: Observations from Tahrir Square', *Journal of Communication* 62.2 (2012): 363–379.

Tukey, J.W. 'Exploratory Data Analysis', 1977, http://xa.yimg.com/kq/groups/16412409/1159714453/name/exploratorydataanalysis.pdf.

Tumasjan, A., T.O. Sprenger, P.G. Sandner, and I.M. Welpe. 'Predicting Elections with Twitter: What 140 Characters Reveal About Political Sentiment', in *Fourth International AAAI Conference on Weblogs and Social Media*, 2010, https://www.aaai.org/ocs/index.php/ICWSM/ICWSM10/paper/view/1441.

Twitter. 'Getting started with Twitter', https://support.twitter.com/articles/215585.

Ungar, S. 'The Rise and (Relative) Decline of Global Warming as a Social Problem', *The Sociological Quarterly* 33.4 (1992): 483–501.

Union of Concerned Scientists. 'Smoke, Mirrors and Hot Air: How ExxonMobil Uses Big Tobacco's Tactics to Manufacture Uncertainty on Climate Change', 2007, http://www.ucsusa.org/sites/default/files/legacy/assets/documents/global_warming/exxon_report.pdf.

United Nations. 'United Nations Framework Convention on Climate Change', 1992, https://unfccc.int/files/essential_background/background_publications_htmlpdf/application/epdf/conveng.pdf.

United Nations Conference on Climate Change, 'COP21', 2015, http://www.cop21.gouv.fr/en/.

United Nations Framework Convention on Climate Change. 'Adoption of the Paris Agreement', United Nations, 12 December 2015, https://unfccc.int/resource/docs/2015/cop21/eng/l09.pdf.

Vaidhyanathan, S. *The Googlization of Everything: (And Why We Should Worry)*, Berkeley, CA: University of California Press, 2011.

Van Dijck, J. 'Tracing Twitter: The Rise of a Microblogging Platform', *International Journal of Media & Cultural Politics* 7.3 (2011): 333–348.

Van Dijck, J. *The Culture of Connectivity: A Critical History of Social Media*, New York, NY: Oxford University Press, 2013.

Veltri, G. 'Microblogging and Nanotweets: Nanotechnology on Twitter', *Public Understanding of Science* 22.7 (2013): 832–849.

Venturini, T. 'Diving in Magma: How to Explore Controversies with Actor-network Theory', *Public Understanding of Science* 19.3 (2009): 258–273.

Venturini, T. 'Building on Faults: How to Represent Controversies with Digital Methods', *Public Understanding of Science* 21.7 (2010): 196–812.

Venturini, T. 'What Is Second-degree Objectivity and How Could It Be Represented', unpublished ms., 2011, http://www.medialab.sciences-po.fr/publications/Venturini-Second_Degree_Objectivity_draft1.pdf.

Venturini, T., A. Meunier, A. Munk, R. Rogers, E. Borra, B. Rieder, et al. 'Climaps by EMAPS in 2 Pages: A Summary for Policy Makers and Busy People', 2014, http://papers.ssrn.com/sol3/papers.cfm?abstract_id=2532946.

Weare, C. and W. Lin. 'Content Analysis of the World Wide Web: Opportunities and Challenges', *Social Science Computer Review* 18 (2010): 272–292.

Weltevrede, E. *Repurposing Digital Methods: The Research Affordances of Platforms and Engines*, University of Amsterdam, Amsterdam, 2016.

Weltevrede, E., A. Helmond, and C. Gerlitz. 'The Politics of Real-time: A Device Perspective on Social Media Platforms and Search Engines', *Theory, Culture & Society* (2014): 0263276414537318, http://doi.org/10.1177/0263276414537318

Wikimedia contributors. 'Wikipedia:Bots/Requests for Approval/SmackBot 0', 3 November 2010, https://en.wikipedia.org/w/index.php?title=Wikipedia:Bots/Requests_for_approval/SmackBot_0&oldid=394496539.

Wikimedia contributors. 'Bot Activity Matrix', http://stats.wikimedia.org/EN/BotActivityMatrix.htm.

Wikimedia contributors. 'Editing Frequency of All Bots', 3 March 2018, http://en.wikipedia.org/wiki/Wikipedia:Editing_frequency/All_bots.

Wikimedia contributors. 'History of Wikipedia Bots', 30 November 2017, http://en.wikipedia.org/wiki/Wikipedia:History_of_Wikipedia_bots.

Wikimedia contributors. 'Lamest Ddit Wars', 17 July 2019, http://en.wikipedia.org/wiki/Wikipedia:Lamest_edit_wars.

Wikimedia contributors. 'List of Wikipedians by Number of Edits,' 15 August 2019, http://en.wikipedia.org/wiki/Wikipedia:List_of_Wikipedians_by_number_of_edits#List.

Wikimedia contributors. 'System Administators', 3 August 2019, http://meta.wikimedia.org/wiki/System_administrators.

Wikimedia contributors. 'User:Ram-Man', 1 March 2016, https://en.wikipedia.org/w/index.php?title=User:Ram-Man&oldid=707772255.

Wikimedia contributors. 'Wikipedia Bot Policy', 4 July 2019, http://en.wikipedia.org/wiki/Wikipedia:Bot_policy.

Wikimedia contributors. 'Wikipedia:Manual of Style/Linking', 5 March 2016, https://en.wikipedia.org/w/index.php?title=Wikipedia:Manual_of_Style/Linking&oldid=708334675.

Wikimedia contributors. 'Wikipedia User Groups', 13 August 2019, http://meta.wikimedia.org/wiki/User_groups.

'Wikimedia Statistics', http://stats.wikimedia.org/.

Wikipedia contributors. 'Free City of Danzig', 14 August 2019, http://en.wikipedia.org/wiki/Free_City_of_Danzig.

Wikipedia contributors. 'Gdańsk', 10 March 2016, https://en.wikipedia.org/w/index.php?title=Gda%C5%84sk&oldid=709411660.

Wikipedia contributors. 'La Grange, Illinois', 27 February 2016, https://en.wikipedia.org/w/index.php?title=La_Grange,_Illinois&oldid=707244890.

Wolters, T. 'Alarmistische KNAW in Grote Problemen', 2011, http://climategate.nl/2011/10/25/alarmistische-knaw-in-grote-problemen/.

Wolters, T. 'Bad Science in Alarmist Report from Royal Dutch Academic Council', 2011, http://climategate.nl/2011/10/19/bad-science-in-alarmist-report-from-royal-dutch-academic-council/.

Wood, D. and J. Fels. *The Power of Maps*, New York: The Guilford Press, 1992.

Wouters, P. *The Citation Culture*, Amsterdam: University of Amsterdam, 1999.

Zittrain, J. *The Future of the Internet*, New York: Penguin, 2008.

ACKNOWLEDGMENTS OF COLLABORATIVE WORK

The research presented in this publication is a culmination of climate change-related projects that I have worked on since 2007 with many colleagues from the Digital Methods Initiative at the University of Amsterdam and in collaboration with our research partners at Médialab Sciences Po, Density Design at Politecnico di Milano, and Digital Sociology at Goldsmith's University in London. This book has known a previous life, in the form of a Ph.D. dissertation with the University of Amsterdam Department of Media Studies. The feedback and suggestions provided by my supervisors, Prof. dr. José van Dijck and dr. Bernhard Rieder, as well as the impeccable proof-reading and copy-editing by Rachel O'Reilly, have proven invaluable.

At the Amsterdam University of Applied Sciences I have worked at the Institute of Network Cultures collaborating with many fellow organizations on international research conferences such as the Society of the Query (on search engine critique) and Video Vortex (on visual media after YouTube), and on data literacy projects with the Citizen Data Lab as part of Amsterdam Creative Industries Network. While it is not easy to disentangle the many ideas and influences traversing these years, I highly value to at least attempt to properly acknowledge these collaborative efforts, which have led to the chapters of this book and refer to the previously published work this has resulted in.

Chapter 1: Introduction

The idea of formulating the research that has led to this book has its origins in a project during the Digital Methods summer school of 2007, the first annual summer program on methods and tools for social research with the web at the University of Amsterdam, titled 'New Objects of Study.' Here, I worked with Esther Weltevrede on our project titled 'Issue Animals', which I briefly discuss in this chapter. The idea of the technicity of web content I publicly presented for the first time at 'Enquiring Minds,' a research seminar as part of *PICNIC08*, 24-26 September 2008, Amsterdam.

The discussion in this chapter of the work of Bruno Latour and Noortje Marres has also been published in a paper co-authored with Ruurd Priester. In that research project, we applied some of the methods proposed in this book to an analysis of bottom-up initiatives in the city of Amsterdam. The paper has been published in a special issue of Computer Supported Cooperative Work.

S. Niederer and R. Priester, 'Smart Citizens: Exploring the Tools of the Urban Bottom-up Movement', *Computer Supported Cooperative Work*, (2016) 25: 137-152, https://doi.orgf/10.1007/s10606-016-9249-6.

Chapter 2: Foundations of Content Analysis

This chapter was written mostly during my time spent as a visiting scholar at the Annenberg School for Communication, University of Pennsylvania. Here I had the opportunity to work with scholars at the core of Content Analysis and collaborate with the Iran Media Group there on a study of Internet censorship, which was published by the Annenberg School and included in the *Blackwell Companion to New Media Dynamics*.

R. Rogers, E. Weltevrede, S. Niederer, and E. Borra, 'National Web Studies: The case of Iran', in J. Hartley, J. Burgess, & A. Bruns (eds) *Blackwell Companion to New Media Dynamics*, Oxford: Blackwell, 2013, pp. 142-166.

Chapter 3: Climate Debate Actors in Science and on the Web

A full version of this chapter in a previous form has been published in the journal *Necsus* (2013). The chapter is a culmination of multiple collaborative projects. The first is a scientometric analysis and extended mapping of climate skeptics, which I conducted during the Digital Methods summer school of 2008 in collaboration with Andrei Mogoutov, developer and owner of ReseauLu, and Bram Nijhof (at the time a student of the New Media & Digital Culture master's program at the University of Amsterdam). The first version of this study was published on the online research platform www.mappingcontroversies.net, as part of the EU 7th Framework project Macospol in 2009, with scientific coordinator Bruno Latour.

My collaboration with the French climate journalist Denis Delbecq led to the mapping of French climate skepticism, which resulted in a co-authored online publication (2010). In 2011, I continued this research with a mapping of Dutch climate skepticism. I have presented this research at *Media in Transition 7* (MiT7) in Cambridge, Massachusetts (May 2011), the *Media of Collective Intelligence* event at the University of Siegen (November 2011) and the conference *Data Traces: Big Data in the Context of Culture and Society* at the Institute of Experimental Design and Media Cultures in Basel (July 2015).

D. Delbecq and S. Niederer, 'Climatosceptiques et Climatologues, Quelle Place Sur l'Internet?', 2010, http://effetsdeterre.fr/2010/04/12/climatosceptiques-quelle-place-sur-linternet/.

S. Niederer, 'Climate Change Skeptics on the Web', 2009, https://web.archive.org/web/20140621023333/http://www.mappingcontroversies.net/Home/PlatformClimate ChangeSkepticsScience.

S. Niederer, 'Global Warming Is Not a Crisis! Studying Climate Change Skepticism on the Web', *Necsus* 3 (Spring, 2013): http://www.necsus-ejms.org/global-warming-is-not-a-crisis-studying-climate-change-skepticism-on-the-web/.

Chapter 4: Wikipedia as Socio-technical Utility for Networked Content Analysis

This chapter is based on research conducted at a Digital Methods Summer School of 2009, with Richard Rogers, Zachary Deveraux, Bram Nijhof, and Auke Touwslager, in which we compared the dependency of Wikipedia on bots for editing, in the various language versions of Wikipedia. In 2010, a discussion of this research with my Ph.D. supervisor José van Dijck led to the decision to develop this research further into a co-authored paper. This paper was published in the journal *New Media & Society* and reprinted in 2014 as part of the edited volume *Researching Society Online.*

The research for this chapter has led to an interview about Wikipedia with Nicholas Nova, which is published in his book *Wikipedia's Miracle* (and in the French edition *Le Miracle Wikipedia*).

I have presented various versions of this chapter at *Formatting Utopia – from Paul Otlet to the Internet*, a conference at the Mundaneum in Mons/Bergen in Belgium (November 2008), ATACD *Changing Cultures, Cultures of Change* conference at the University of Barcelona (December 2009), as a *Brown Bag Lecture* at the Amsterdam University of Applied Sciences (January 2010), at *Medien der Kollektiven Intelligenz* at the University of Konstanz, at the Annenberg School for Communication in Philadelphia (April 2011), at the *ReClaiming Participation* conference in Zürich (May 2014) and at *Reading Wikipedia,* the Praemium Erasmianum Conference at the Royal Netherlands Academy of Arts and Sciences (KNAW) in November of 2015.

S. Niederer, 'Interview', in N. Nova and F. Kaplan (eds) *Wikipedia's Miracle*, Lausanne: EPFL Press, 2016, pp. 53-61.

S. Niederer, 'Entretien', in N. Nova and F. Kaplan (eds) *Le Miracle Wikipedia,* Lausanne: PPUR, 2016, pp. 51-61. (French edition)

S. Niederer and J. van Dijck, 'Wisdom of the Crowd or Technicity of Content? Wikipedia as a Sociotechnical System', *New Media & Society* 12.8 (2010): 1368–1387.

S. Niederer and J. van Dijck, 'Wisdom of the Crowd or Technicity of Content? Wikipedia as a Sociotechnical System', in M. David and P. Milward (eds) *Researching Society Online*, London: Sage, 2014 (Reprint).

Chapter 5: Mapping the Resonance of Climate Change Discourses in Twitter

This study was published in the online publication *Climaps: An Online Issue-Atlas of Climate Change Adaptation* and featured in *Wired Italia.* The United Nations tweeted the study's climate vulnerability world map after its publication in the *Climaps* atlas.

For the case studies in this chapter, which were conducted in data sprints as part of the EU FP7 project Electronic Maps to Assist Public Science (EMAPS), I have collaborated closely with researchers at the Digital Methods Initiative, including Sophie Waterloo, Simeona Petkova, Natalia Sanchez Quérubin, Liliana Bounegru, and Catalina Iorga. Erik Borra and Bernhard Rieder are the developers of the tools used for this analysis. The research team also consisted of information designers from DensityDesign in Milan, including Carlo de Gaetano, Gabriele Colombo, and Stefania Guerra.

After the data sprint, the descriptions of our various case studies, which I worked on with various team members but especially Sophie Waterloo and Gabriele Colombo, were compiled and expertly edited by Natalia Sanchez Quérubin and Lilliana Bounegru for publication in the online issue atlas *Climaps.eu*, which presents the issue stories and issue maps of the EMAPS project. For this chapter, I have re-assembled and subsequently rewritten the Climaps materials and our original descriptions (co-authored with Sophie Waterloo and Gabriele Colombo) to suit the focus of this book and aptly present the collaborative case studies of mapping the climate debate with Twitter. I have presented the research at the conference *Social Media and the Transformation of Public Space,* at the Royal Netherlands Academy of the Arts and Sciences (June 2014).

R. Battaglia, 'Clima, Ecco la Mappa di Chi Litiga Sul Cambiamento Climatico', *Wired Italia*, 2014, http://www.wired.it/attualita/politica/2014/12/15/emaps-come-mappare-disaccordo-clima/.

EMAPS, 'Profiling Adaptation and Its Place In Climate Change Debates With Twitter', 2014, http://climaps.eu/#!/map/profiling-adaptation-and-its-place-in-climate-change-debates-with-twitter-I.

EMAPS, 'Reading the State of Climate Change From the Web: Top Google Results', 2014, http://climaps.eu/#!/map/profiling-adaptation-and-its-place-in-climate-change-debates-with-twitter-ii.

EMAPS, 'Who is Vulnerable According to Whom?', 2014, http://climaps.eu/#!/map/who-is-vulnerable-according-to-whom.

@UNEnvironementUNEP, 'This map compares 3 lists of countries ranked by #climeatechange vulnerability. Surprised? buff.ly/1tFvIZJ', Twitter post, 1 November 2014, 3:12 PM, https://twitter.com/unep/status/528550060397957120.

Chapter 6: Conclusions

In the conclusions, I mention three studies that I have worked on collaboratively: *People's Dashboard, The City as Interface*, and the *Knowledge Mile Atlas*.

The People's Dashboard was a project developed during the Digital Methods Winter School of January 2015. Esther Weltevrede and I facilitated the group, which included the following participants: Evelien Christiaanse, Caio Domingues, Yvette Ducaneaux, Inte Gloerich, Alex Harrison, Hendrik Lehmann, Gabriel Reis, Pavel Rodin, Jurij Smrke, Janina Sommerlad. Erik Borra developed the plugin. Stefania Guerra and Tommaso Renzini (Density Design) made the design.

The City as Interface was a project developed during the summer school of 2015, in which we worked with subject matter expert Martijn de Waal, author of the book *The City as Interface* (2014). The project team, which I facilitated, had as participants Nataliya Tkachenko, Xinyang

Xie (Yang), Peta Mitchell, Maarten Groen, Adrian Bertoli, Khwezi Magwaza, Naomi Bueno de Mesquita, Joe Shaw, Alexander van Someren, Tim Leunissen, and Philip Schuette. Designers working with us on the project were Michele Mauri and Donato Ricci.

The *Knowledge Mile Atlas* is an ongoing collaboration (2014-) with information designers Gabriele Colombo, Michele Mauri, and Matteo Azzi of Density Design and Calibro, and Carlo De Gaetano, Federica Bardelli, Wouter Meys, Maarten Groen, Maarten Terpstra, and Matthijs ten Berge at the Amsterdam University of Applied Sciences, in which we analyze and visualize the online presence and resonance of an urban area under development. The area cuts through the city center of Amsterdam and crosses many districts and neighborhood borders. We presented our co-authored paper about the Knowledge Mile maps at the conference *Hybrid City 3: Data to the People* (September 2015) at the University of Athens. The paper was published in the conference proceedings.

The People's Dashboard is described extensively on the wiki project page: https://wiki.digitalmethods.net/Dmi/PeoplesDashboard. To install the plugin, go to: https://github.com/digitalmethodsinitiative/peoplesdashboard.

The wiki page for the project *The City as Interface* is: https://wiki.digitalmethods.net/Dmi/TheCityAsInterface.

S. Niederer, G. Colombo, M. Mauri, and M. Azzi, 'Street-Level City Analytics: Mapping the Amsterdam Knowledge Mile,' in *Hybrid City 3: Data to the People*, Athens: University of Athens, 2015, www.media.uoa.gr/hybridcity.

A previous version of this chapter has been published as:

S. Niederer, 'The Study of Networked Content: Five Considerations for Digital Research in the Humanities', in G. Schiuma and D. Carlucci (eds) *Big Data in the Arts and Humanities*, New York: Auerbach Publications, 2018, pp. 89-100, https://doi.org/10.1201/b19744